A HISTORY
OF THE CIRCLE

A HISTORY
OF THE CIRCLE

Mathematical Reasoning and the Physical Universe

ERNEST ZEBROWSKI, JR.

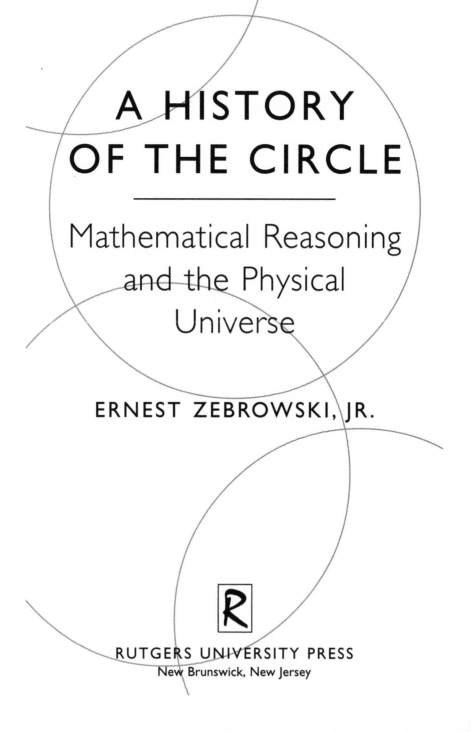

RUTGERS UNIVERSITY PRESS

New Brunswick, New Jersey

Zebrowski, Ernest.
    A history of the circle : mathematical reasoning and the physical
universe / Ernest Zebrowski.
        p.   cm.
    Includes bibliographical reference and index.
    ISBN 0-8135-2677-9 (cloth : alk. paper)
    1. Cycles.   2. Pi.   3. Circle.   I. Title.
    Q176.Z42   1999   98-37636
    510—dc21   CIP

Manufactured in the United States of America

This book is dedicated
to the memory of

Josefa V. Zajac Zebrowski
1916–1995

and

Ernest T. Zebrowski
1916–1995

CONTENTS

PREFACE

This book is not a history in the conventional sense. It is, rather, an exploration of the mysterious fact that natural phenomena behave in ways that allow us to link the past with the future. Nor is this book meant to be an exhaustive treatise on the mathematics of the circle. I use the circle as an exemplar that allows us to examine the broader relationship between mathematical reasoning and the physical universe. In particular, the following question begs to be examined: How is it that mathematics, ostensibly an abstract creation of the human mind, can produce conclusions that seem to be valid in the external realm of *physical* reality? This book is my humble attempt to address that question.

I write, not to practicing mathematicians and scientific theorists, but rather to the broader community of educators, students, and independent thinkers who seek to make sense of scientific "truths" that various specialists have discovered through mathematical logic. My theme does not require that the reader delve very deeply into the mathematics itself. Although I have included some examples in the symbolic formalisms of algebra and trigonometry, it is not essential that the reader be an expert in these areas, and skipping over some of the equations will do little injustice to my discussions. This book is intended to be read, not computed.

Each of us has experienced those moments in life when we are overcome by a feeling of deep intimacy with nature. For some, it happens in hiking through a remote wilderness; for others a gentle rain in the city will do. For me, it always happens when I find myself alone on a beach. It was on such an occasion that the idea for this book surfed in on the waves.

During my quarter-century of college teaching, I've presented the theory of wave motion, at different levels, to around seven or eight thousand students. Invariably what excites them most are my digressions from the pure physics— my commenting on historical incidents of rogue waves, great tsunamis, and giant storm swells. I'd sometimes felt a bit uneasy about digressing in this manner, talking about incidents that did not lay foundations for either the physics or the mathematics, but which instead were curious historical events that were fun to talk about. This particular day, my feet in the ocean, it dawned on me that no professor *ever* teaches anything but history, regardless of the discipline or topic. We teach what has been learned in the past, hoping that the universe will cooperate in allowing our students to apply this past knowledge to the unfolding future. This is as true in music and art as it is in science and mathematics. We could not teach, nor could anyone learn, if we did not believe that the universe displays a historical continuity.

Clearly, a mathematical approach to predicting future natural events is more reliable than alternatives such as horoscopes or palm-reading. Yet mathematical reasoning itself is not science. Science never uncovers absolute truths, because it is impossible to observe the entire universe for all time; what we observe in nature is only an infinitesimal slice of what has been or can be. Yet from a few paltry centuries of observation of extremely local events, our modern science seeks to draw grand conclusions about the master design of the universe. Clearly, science is based on plausible reasoning, not the demonstrative reasoning of mathematics. Yet, equally clearly, the two are intimately and profoundly linked.

As for the waves at the beach: in the distance they are a chaotic chop; closer to shore, they rise as a series of undulating crests and troughs; and at the surf line they explode into chaos once more. Equations can be written to describe, at least approximately, all of these effects. And ultimately, such equations turn out to be based on the geometry of the circle. But do we actually *see* such circles when we watch the surf? Yes, if we pay attention closely enough, we not only see the circles, but sometimes we can actually feel them.

The waves at the beach, the chariot wheels of the ancient Egyptians, the swirling of millions of stars in distant galaxies, the shapes of puffball mushrooms, the structure of the atoms in our own bodies—all these and many other diverse physical entities become connected when we reduce their descriptions to mathematical language. But are the connections really there, and historically valid, or are they just artifacts of our limited way of thinking mathematically? I invite you to take this little journey with me and draw your own conclusions.

My thanks to Ian Scott, who prepared the artwork, and Paul Craig, who made helpful suggestions on the content.

Ernest Zebrowski, Jr.
Southern University
Baton Rouge, Louisiana
November 1998

A HISTORY
OF THE CIRCLE

CHAPTER 1

THE QUEST FOR PI

This book is about something that doesn't physically exist. Try to find a circle in nature, or even in the artifacts of human civilization, and you've embarked on an impossible hunt. A car tire? No, every tire is a bit flat on the bottom. A quarter? No, quarters have little ridges around their rims. A doughnut? Too bumpy, if we look closely. Even the full moon has mountains and craters around its perimeter, and we need only a decent pair of binoculars to see them.

But suppose we take out a compass and a piece of paper, and draw what appears to be a circle. Isn't this a *true* circle? No. A pencil line can never have a perfectly uniform thickness, and there are other mechanical reasons why a "circle" drawn with a compass has minor variations in its radius. Once again, if we look closely, say through a magnifying glass, we notice that our curve has a fuzzy edge. A true circle is only a figment of the human imagination. Even a computer can't draw one.

But isn't this just an abstract philosophical issue? Surely, a physical circle is often *close enough* to a true circle that for all practical purposes we can consider it to be one. Indeed, if this weren't the case, a whole host of devices we take for granted would not be possible, from bottle caps to automobile transmissions. Why concern ourselves about the nonexistence of true circles, when real but approximate circles seem to serve our human needs so well?

The answer, which forms a thread running through this book, is that approximate circles *don't* always serve our human needs, nor are shapes that appear to be approximately circular always best described as mathematical circles. The question How round is round enough? it turns out, does not have a simple answer, and deciding when the geometry of "perfect" circles has something to say about real physical objects has historically often presented considerable intellectual challenges. Do we fool ourselves in any important way, for instance, if we say that our planet Earth travels in a circle around the sun? (Johannes Kepler thought so, and in 1609 he triggered a scientific revolution as a result.) Do architects miss any opportunities if they build structural arches in the shape of geometrical semicircles, as did the Romans? (Arab and European builders of c. 1000 C.E. thought so, and their new insights ushered in a golden age of mosque and cathedral building.) If we seek to understand the workings of nature, How round is round enough? is not a trivial question.

This book is an odyssey through two worlds: one the world of round physical entities, the other the world of true circles and their mathematical cousins. At points where these two worlds intersect, we'll use the opportunity to step from one to the other. Along the way, we'll explore some of the marvels of ancient and modern engineering, we'll examine some principles of architecture and art, we'll dwell on a few questions of geography and astronomy, and we'll consider questions of the grand design of the universe. We'll look at some ancient abstract ideas that have stood the test of time, and some more modern ideas that have failed. Throughout, the question How round is round enough? will remain one of our guides.

The Constant Circle Ratio

The concept of a *ratio* is a very old one, dating back to civilization's earliest written records. If it takes an acre of pasture to graze two sheep, then the ratio of pasture to sheep is 1 to 2 (alternatively written as 1:2 or ½). In itself this is just a definition, and not particularly interesting. What *is* interesting, though, is that more pasture is needed for more sheep in approximately this same ratio of 1 to 2, so 250 acres of pasture (for instance) can be expected to support about 500 sheep. An ancient sheep herder who recognized the constancy of such a ratio had the ability to make predictions, predictions that both prepared and empowered that person.

With the rise of cities in ancient Egypt, Sumeria, and Mesopotamia, it became important to be able to predict the collective future needs of the population. The coming demand for food could be computed by using the concept of the ratio, and similar ratio-based computations could also predict future tax income to the royal treasury. In fact, of the hundreds of tons of ancient Mesopotamian clay tablets unearthed by archaeologists, most contain nothing more than commercial transactions, inventories, and ratio calculations.

Yet there is nothing in nature's rule book that requires every ratio to be constant. Most ratios, in fact, are *not* constant. If, for instance, it took 24 rowers to row a galley at 15 mi/h, this does not mean that 48 rowers would get the boat up to 30 mi/h and that with 144 rowers the boat would hit 90 mi/h. (In fact, this line of reasoning would suggest that the ancients could have broken the sound barrier just by getting together enough rowers.) If it took 1,000 stone blocks to build a pyramid 20 meters tall, this does not mean that 2,000 identical blocks would build a pyramid 40 meters tall. (In fact, we will see in chapter 6 why doubling the number of blocks increases a pyramid's height by only about 26%.) Although it's a simple matter for an accountant or mathematician to assert that a particular ratio is constant, natural law is the final arbiter. Clearly, before making predictions on the basis of an assumed constant ratio, we first need to get someone to check out the reality of the situation.

The wheel and its precursors (the roller and the pulley) were important ancient inventions, and their widespread adoption led to the following question: If a wheel rolls through one complete revolution without slipping on its supporting surface, how far forward does it roll? Clearly, big wheels go farther than smaller wheels, but is there any constant ratio between a wheel's diameter and its forward movement that permits a prediction? Does a wheel's forward movement behave like the relatively constant ratio of pasture to sheep, or more like the variable ratio of a boat's speed to the number of rowers?

Undoubtedly, the earliest answers were provided by measurement. Imagine an early wheel, which probably wasn't very round, and which rolled on a surface that wasn't quite flat. By marking a starting point and an end point for one revolution, it could be established that the wheel advanced forward a little over three diameters when it rolled through one revolution (Fig. 1.1). No sophisticated measuring equipment was needed to discover this; it would simply be a matter of knotting a piece of rope to record the wheel's diameter, and counting how many of these diameters

3

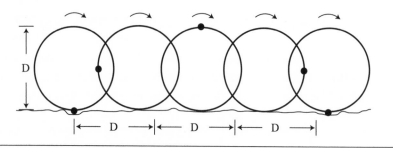

Figure 1.1. A rolling wheel of diameter D. In one revolution, the wheel travels a distance of slightly more than three diameters.

one needed to lay down on the ground to add up to one revolution. The result:

$$\frac{1 \text{ revolution}}{1 \text{ diameter}} = 3 \text{ (approximately).}$$

Doing this once, of course, proves nothing. Doing it dozens of times with wheels of different sizes begins to suggest a pattern. Eventually, when many different people make such measurements independently, evidence accumulates that the ratio of the circumference to the diameter of a wheel is indeed a constant, equal to a little more than 3, regardless of the physical size of the wheel. And clearly, if we keep the wheel in place and instead drape a rope around it, the ratio of the distance around to the distance across has the same value of a little more than 3.[1]

Given this preliminary suggestion of a constant ratio, it becomes possible to do a more precise experiment: draw the outline of a large wheel in the sand, or with charcoal on a big flat slab of marble, and forget about the complications of using a physical wheel. By 2000 B.C.E., the Babylonians took this approach and experimentally established the constant circle ratio as 3⅛, or in modern decimal notation, 3.125. The ancient Egyptians arrived at a slightly different value, usually quoted as 3⅐, or about 3.143 (although during at least some periods of Egyptian history the "official" figure seems to have been slightly higher). Which of these values one chose to use mattered little in those early days, because the main thing the constant circle ratio applied to was the wheel, and ancient wheels weren't all that round.

The important discovery was that circular shapes do indeed have a relatively constant ratio of circumference to diameter. So, for instance, if a wheel had a diameter of 2 cubits (2 forearm-lengths), and it rolled through 1,000 revolutions, one could predict it would advance a distance of $2 \times 3\frac{1}{7} \times 1{,}000$, or about 6,286 cubits, more or less. If a second wheel had twice the diameter, it would travel twice as far in the same number of revolutions. Half the diameter, and it would travel half as far. And so on.

Calculating π

Around 1000 C.E., Arab mathematicians adopted a decimal system for representing numbers, and over the next few centuries this notational system filtered into the West. The equal sign appeared in 1557, and by the 1700s it was standard practice to symbolize the constant circle ratio by π (the Greek letter "pi"), where

$$\pi = \text{circumference/diameter.}$$

Historical records tell us that attempts to find an "exact" value of π occupied a great many thinkers through the ages. Some of the results we know about are summarized in Table 1.1.[2] Today, the value of π has been computed (although not printed out) to over a billion decimal places, and modern computers can easily extend the string of digits without limit. No matter how many digits we calculate, however, we can be guaranteed we'll never reach π's last digit. These digits turn out to continue infinitely, in a nonrepeating sequence. The only practical application for knowing π to millions of decimal places is that it is useful as a test for new computers and their software, providing a benchmark for assessing both speed and accuracy.

Clearly, none of the estimates of π in Table 1.1 could have been arrived at by rolling wheels on pavements, or even by drawing circles and measuring them. Suppose, for instance, that we can measure lengths to a precision of 1 millimeter (which would have been extremely good in ancient times), and that we want to measure π with a precision to the third decimal place, that is, 3.142. How big a circle do we need to draw? One, it turns out, on the order of 10 meters in diameter, or about 33 feet![3] For more precision, we'd need even bigger circles, and we'd quickly run into all sorts of practical problems, like whether the surface is truly flat, whether the measuring instrument itself has irregularities, and whether the "circle" we've drawn is really as round as we need it to be.

Table 1.1. Historical progress in determining the value of π.

Mathematician	Date	Value of π
Archimedes of Syracuse	250 B.C.E.	$3.1408 < \pi < 3.1429$
		$\sim 211875 : 67441 = \sim 3.14163$
Hou Han Shu (China)	c. 130 C.E.	~ 3.1622
Ptolemy of Alexandria	c. 150	$377 : 120 = \sim 3.14167$
Liu Hui (China)	264	$3.141024 < \pi < 3.142704$
		~ 3.14159
Tsu Chung-Chih (China)	c. 400	$3.1415926 < \pi < 3.1415927$
Aryabhata (India)	499	$62,832 : 20,000 = 3.1416$
Fibonacci (Italy)	1202	$864 : 275 = \sim 3.141818$
Al-Kashi (India)	c. 1436	3.141 592 653 589 79
Viète (France)	1593	3.141 592 653 6
Van Ceulen (Holland)	1596	32 decimal places
De Lagny (France)	1717	127 decimal places
Vega (Spain)	1794	140 decimal places
Dase (Germany)	1844	200 decimal places
Richter (Germany)	1855	500 decimal places
Ferguson (England)	1947	808 decimal places
Shanks & Wrench (U.S.)	1961	100,265 decimal places
Current computers	1998	> 50 billion decimal places

Archimedes of Syracuse (c. 250 B.C.E.) fully understood these practical limitations on any experimental approach to finding π. As an alternative, he began with the observation that in drawing regular polygons of increasing numbers of sides, the shapes begin to look more and more like a circle (Fig. 1.2). From this perspective, a circle is a regular polygon with a huge number of very short straight sides, a conceptual leap that rendered a whole new approach to determining π.

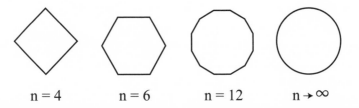

$n = 4$ $n = 6$ $n = 12$ $n \rightarrow \infty$

Figure 1.2. Increasing the number of sides of a regular polygon makes the shape more closely approximate a circle.

The advantage in visualizing a circle as a many-sided polygon is that it is relatively easy to calculate the perimeter of a regular polygon. Unfortunately, the calculation can also be very tedious, and Archimedes lacked not only a calculator but also a decimal notation. Million-sided or even thousand-sided polygons were clearly out of the question for practical computational reasons. Yet clearly a four- or six-sided polygon fell short of being round enough. For Archimedes, the question of how round would be round enough reduced to a judgment call; ultimately, he decided that a 96-sided polygon was close enough to a circle that the perimeter of such a polygon would be a reasonable approximation to the circumference of a circle. And, in fact, if you build a 96-sided polygon and roll it on a flat surface, it will indeed roll along reasonably well.

Using this line of logic, Archimedes calculated the value of π correctly as far as 3.1416, and he also established limits on how far the "true" value could possibly deviate from this estimate. To find these limits, Archimedes calculated a pair of perimeters: that of a 96-sided polygon that just fit inside a unit circle, and a second for a 96-sided polygon that snugly enclosed a unit circle. Figure 1.3 shows how this calculation works for the simpler (but much more approximate) case of a six-sided polygon. Using 96 sides, Archimedes ultimately decided that π could not possibly be any larger than 3.1429, nor any smaller than 3.1408.

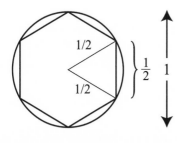

Each side of this hexagon has a length equal to the circle's radius, or $\frac{1}{2}$. The perimeter of the hexagon is $6(\frac{1}{2}) = 3$.

Each side of this hexagon has a length of $\sqrt{\frac{3}{3}}$. The perimeter of the hexagon is $6(\sqrt{\frac{3}{3}}) = 3.46$.

Figure 1.3. Archimedes' polygon method for finding the upper and lower limits of π. The inscribed and circumscribed hexagons shown above establish that $3 < \pi \leq 3.46$. Archimedes used 96-sided polygons to find that $3.1408 < \pi < 3.1429$.

Archimedes died in Syracuse in 212 B.C.E., when a horde of Roman soldiers burst into the palace and found him squatting on the marble floor, working out a geometrical theorem in chalk. Archimedes jumped up and demanded that the invaders not disturb his circles. Unfortunately, his shouts in the Greek language were misunderstood by the Roman ears, and although the Roman general Marcus Marcellus had ordered that he be captured alive, Archimedes was killed on the spot. He died surrounded by his circles.[4]

The computational legacy of Archimedes can be summarized as follows: One can approximate a circle as a regular polygon, and the more sides to the polygon, the better the approximation. In the limit of an infinite number of sides of infinitesimal length, all geometrical features of a true regular polygon are identical to the corresponding features of a true circle. Computing π (or proving theorems) based on this scheme can be done in the abstract, avoiding the limitations of measuring real polygons and circular shapes that have inherent irregularities. Because conclusions are based on mathematical logic, with drawings and sketches serving only as a guide, irregularities in drawings have no detrimental effect on the accuracy of the mathematical result. As for circles in the physical world, which are always imperfect, we can logically assume that the closer they come to true circles, the closer will their ratios of circumference to diameter approach π.

For the next thousand years, many other mathematicians (some of whom knew nothing of Archimedes) used polygons of various numbers of sides to calculate π. Embarking on such a calculation meant accepting from the outset that the result would not be exact, for unless a polygon has an infinite number of sides, it is still not a circle. More limiting still, one needed to decide ahead of time how many sides to use, which in turn established the limit on the final precision. It is not possible, using the polygon method, to decide to improve on the precision once the calculation has commenced.

Clearly, then, both the experimental method and the polygon method have serious shortcomings. Is there an alternative method for calculating π? Yes. In 1671, a Scotsman, James Gregory, discovered that π can be calculated without even dealing with polygons, but rather by summing the terms of the following infinite series:

$$\pi = 4\left(1 - \tfrac{1}{3} + \tfrac{1}{5} - \tfrac{1}{7} + \tfrac{1}{9} - \tfrac{1}{11} + - \cdots\right).$$

It turns out that the fractions in this series shrink sufficiently quickly that their algebraic sum converges toward a fixed value, which happens to be π.[5] Unfortunately, though, this particular series is not an expedient method for finding π, because even after three hundred fractions are summed in this manner, the answer is still accurate to only two decimal places, 3.14. Such a level of precision had already been established by Archimedes a thousand years before Gregory came along.

Gregory's discovery was nevertheless important in that he had demonstrated the feasibility of computing π by summing a series of fractions, without any reference to polygons and without having to decide ahead of time "how round" would be "round enough." This spurred others to look for different numerical series that might converge more rapidly, and sure enough, some diligent mathematicians began to find them. One calculation, usually attributed to Isaac Newton, is:

$$\pi = 6[\tfrac{1}{2} + 1/(2 \cdot 3 \cdot 2^3) + (1 \cdot 3)/(2 \cdot 4 \cdot 5 \cdot 2^5) + (1 \cdot 3 \cdot 5)/(2 \cdot 4 \cdot 6 \cdot 7 \cdot 2^7)$$
$$+ (1 \cdot 3 \cdot 5 \cdot 7)/(2 \cdot 4 \cdot 6 \cdot 8 \cdot 9 \cdot 2^9) + (1 \cdot 3 \cdot 5 \cdot 7 \cdot 9)/(2 \cdot 4 \cdot 6 \cdot 8 \cdot 10 \cdot 11 \cdot 2^{11}) + \cdots].$$

We can verify with a calculator that summing just the first four terms leads to 3.14115. Summing the first twenty-two terms of this same series gives a value of π correct to sixteen decimal places, which is certainly a tremendous improvement in precision over Archimedes' polygon method.

If time were no object, it wouldn't matter which series were used to compute π, provided that the sum indeed converged to π. As a practical matter, however, it made sense historically to look for a series that converges rapidly. As we've seen above, Gregory's series requires three hundred terms to achieve the same precision that Newton's series provides with just four terms. Clearly, of these two examples, Newton's was the preferred method for computing π, because its rapid convergence resulted in much better precision for the same time spent calculating. Other such series, somewhat clumsier to write, converge even more rapidly than Newton's. This said, I'll spare the reader the details.

And what is the accepted value of π? If we consult a common ten-digit calculator, the value it displays is

$$\pi = 3.141592654.$$

Although this is only the beginning of an infinite string of digits that do not repeat in any recognizable pattern, this approximation is still more precise than most of us could possibly have any use for. Suppose, for instance, that we want to calculate the circumference of Earth at the equator, and that we use the above value of π, which introduces a rounding error in the ninth decimal place. How much error do we get in the planet's circumference because of this rounding error in π? No more than one-quarter inch in 25,000 miles! This degree of precision, based on just ten digits of π, far exceeds the physical reality of the roundness of planet Earth. The next 50 billion known digits of π obviously exceed by a stupendous margin the precision that would be demanded in any practical computation of an observable circumference.

The Irrationality of π

Pi is what mathematicians refer to as an "irrational" number. This unfortunate terminology (which traces its origin to the ancient Greek mathematicians) simply means that π can't be expressed exactly as the ratio of two integers. Using our modern decimal notation, it means that the digits of π neither terminate nor repeat. Notice, for instance,

$$\tfrac{1}{3} = 0.333333333333333333333333333333\ldots \text{ (repeats),}$$
$$\tfrac{1}{2} = 0.500000000000000000000000000000\ldots \text{ (terminates),}$$
$$\tfrac{5}{8} = 0.625000000000000000000000000000\ldots \text{ (terminates),}$$
$$\tfrac{5}{7} = 0.714285714285714285714285714285142\ldots \text{ (repeats),}$$
$$\tfrac{8}{11} = 0.727272727272727272727272727272\ldots \text{ (repeats),}$$
$$\tfrac{7}{9} = 0.777777777777777777777777777777\ldots \text{ (repeats),}$$
$$\tfrac{10}{13} = 0.769230769230769230769230769230692\ldots \text{ (repeats),}$$

but $\pi = 3.14159265358979323846264333832\ldots$ (neither terminates nor repeats).

Now π is hardly the only irrational number. In fact there are an infinity of irrationals, only a few of which are represented by special symbols such as

$$\sqrt{2} = 1.4142135623730950488\ldots$$
$$e = 2.7182818284590452353\ldots$$
$$\text{or } \log_e 10 = 2.3025850929940456840\ldots$$

Meanwhile, all empirical evidence suggests that the results of physical measurements are intrinsically irrational. Each time we increase the precision of any measurement, the new digits we discover add to a string that does not terminate or repeat. A vast amount of experimental data spanning several centuries confirms this contention, and there is no empirical evidence to repudiate it. Even quantum numbers, which by theoretical definition are integers or rational fractions, are meaningful only as rational multipliers of measured and irrational physical constants. I will return to the issue of measurement in chapter 4; for now, my point is that scientific inquiry rides on a thruway paved mostly by irrational numbers.

What all irrational numbers have in common, besides never ending and never falling into a repeating pattern, is that if we examine a few thousand of their digits, about 10% of these digits are zeros, 10% are ones, another 10% are twos, and so on. This is true for every individual irrational number, and it is also true if we mix a few thousand digits of several irrational numbers in a bucket and draw them out at random. There is no way to distinguish one irrational number from another just by looking at frequency counts or other patterns in their digits, because there is always a 10% probability of drawing any particular digit. On the other hand, if we take two *rational* numbers, say $\frac{3}{7}$ (0.428571428571 . . .) and $\frac{4}{11}$ (0.3636363636 . . .), and mix up a few hundred of their digits, we do *not* find 10% zeros, 10% ones, 10% twos, and so on. Instead, this particular mix results in no zeros or nines, 25% threes and sixes, and 8.3% of each of the remaining digits. A clever mathematician could therefore look at such a statistical distribution of digits and either reconstruct the original rational fractions or else reconstruct a limited number of alternative possibilities.

What distinguishes π from $\sqrt{2}$, then, does not lie in their actual patterns of digits. The important distinction is the fundamental definitions: π is the ratio of the distance around a circle to the distance across, while $\sqrt{2}$ is the ratio of the diagonal of a square to the length of its side. The power of working with irrational numbers does not derive from their specific numerical values, which can never be expressed exactly anyway, but rather from what these numbers *mean*. By transferring such meaning from abstract definitions to the real world of physical objects and processes, we can often gain tremendous insights into the workings of nature.

The Implications of π

It matters little that there is no such thing as a circle in the physical world. What is important is this: *As the geometry of any physical object more closely approximates a true circle, the ratio of its circumference to its diameter more closely approximates π.* Conversely, if we make some physical measurements and repeatedly get results that are an approximate multiple of π, this suggests that there is something of roughly circular geometry involved in the underlying physical mechanism.

As a simple example, consider the following: We set out to repair an old fence on a farm, and in taking inventory of the materials we'll need, we measure the distances between fenceposts. These measurements reveal that the old fenceposts are spaced fairly equally, 9 feet 5 inches from one another. Expressed in decimal form, this is 9.417 feet, which strikes us as an odd distance for anyone to deliberately space fenceposts. Why wouldn't the fence builder have chosen a spacing easier to work with: say, an even 8 feet or 10 feet? If we pull out a calculator, however, we may discover that $9.417 \div 3 = 3.14$, which is a close approximation to π. So, can we come up with any reasonable explanation for why someone might want to set a series of fenceposts at a spacing of 3π feet?

One hypothesis is this: It's an inconvenience to drag a measuring tape along the ground to mark the points for digging a series of postholes. Undoubtedly, the original builder was already pulling a small wagon (or possibly pushing a wheelbarrow) that carried his tools and materials. Why not just strike a chalkmark on the side of one of the wheels, and place a fencepost every third complete revolution? If a wheel has a diameter of one foot, then three revolutions move the wheel forward a distance of 3π feet, or about 9 feet 5 inches, and this becomes a kind of natural spacing for the postholes. No measuring tape needed, no extra work required. Of course this arithmetic doesn't actually prove anything, for the agreement between measurement and calculation may still be just a coincidence. Yet, the hypothesis that where π appears a circle lurks hidden does suggest a way of making sense of observations that might otherwise remain senseless.

The preceding example isn't totally artificial. More than a few speculative articles have claimed that the linear dimensions of various ancient ruins are an integer multiple of π when expressed in the prevailing measurement unit of the time. Sometimes the writer goes on to claim that π must have carried some mystical significance for that civilization, for why else would ancient builders have incorporated its value into the dimensions of temples, tombs, and pyramids? Clearly, however, there is an alternative hypothesis that has nothing to do with mysticism. It is simply this: One simple way to measure a linear distance is to roll a wheel along the ground and to count revolutions. When this procedure is used, π enters the linear measurement automatically.

CHAPTER 2

ROLLERS, WHEELS, AND BEARINGS

Many advanced ancient civilizations, including those in the pre-Columbian Americas, found little use for the wheel. Never, however, has any group of people managed to engage in large construction projects without the help of the roller. From the ancient Egyptians to the Aztecs and Incas to the isolated inhabitants of Easter Island in the South Pacific, early civilizations around the world independently learned that rollers can greatly facilitate the transportation of heavy loads. In fact, without the roller, it is inconceivable that any society could have considered erecting great pyramids, temples, and giant stone statues that required the movement of heavy stones over distances of many miles.

The roller principle may seem so simple as to merit little discussion. Find a few straight tree trunks of the same diameter, lay them parallel on the ground, place a heavy load on top, and push. The tree trunks roll and the load moves—much more easily than if it were sliding along the ground. Yet this isn't the whole story, for if we experiment with this arrangement (say by placing a cereal box on some soda straws), we quickly notice that the load always moves farther than the rollers. In fact, as a roller rolls through one revolution, advancing forward a distance of π times its diameter, the load it supports moves forward a distance of 2π times the roller diameter. This process is shown in Figure 2.1. For any load moved on rollers, regardless of the roller diameter, we find the following constant ratio:

$$\frac{\text{forward movement of load}}{\text{forward movement of roller}} = 2.$$

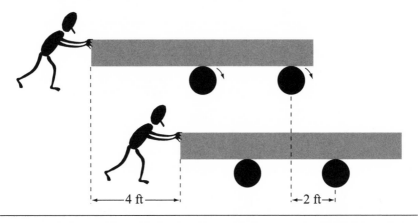

Figure 2.1. Moving a heavy load with rollers. The forward movement of the load is twice the forward movement of the rollers.

This 2:1 motion ratio creates a practical difficulty: it's quite easy to roll the load right off the rollers! The only way to prevent such an unhappy event is to support the load on many rollers, and assign someone to pick up the rollers left behind and carry them forward to reinsert them under the front of the load. Of course, simultaneously, other laborers are needed to push or pull the load. Clearly, the success of such an operation requires coordinated teamwork.

And what is the advantage of this complicated process, compared to simply dragging the load along the ground? A tremendous savings in labor. For instance, to slide a 1,000-lb stone block on a relatively flat hard surface requires a force of 500 lb to 700 lb (and a bit more to initiate the motion). On the other hand, if this same 1,000-lb block is supported by rollers, the effort required to move it may drop to only around 50 lb. By moving a load on rollers rather than dragging it, the required effort may be reduced by as much as 90%.

The effect is dramatic. A number of years ago, two friends and I stopped out of curiousity at a printing shop that was going out of business, and the elderly owner asked if we could move one of his printing presses 80 feet into an adjacent loading area. This particular press had been built around 1895, and its frame was cast iron; it clearly weighed a ton or more. We found a couple of small hydraulic jacks and some sections of iron pipe (old shops usually had items like that lying around); then we raised the press one end at a time and slipped the pipes beneath it. That was the hard part.

Pushing this heavy machine some 80 feet, as it rolled along on the pipe sections, was the easy part. I vividly remember our sense of accomplishment at having done this without working up so much as a drop of sweat.

I don't claim, however, that this incident proves I'd have made a good ancient Egyptian. The ancients, in moving heavy loads great distances, ran into other problems that didn't confront my friends and me in moving the printing press. First, lacking iron pipes, they had the problem of finding suitable rollers. Not just any tree trunk would do. Not only must a roller be fairly round, but if it has even a slight taper, say a 10% change in diameter over the length that supports the load, each revolution will propel one edge of the load 10% farther than the other. In this case, the load does not move forward in a straight line, but instead veers to the right or left. Using tree trunks as rollers, this effect is unavoidable. It can be dealt with only by correction; that is, if the load is swerving to the left, the fat end of the next roller is inserted under the left side of the load so the motion will veer back toward the right. As a result, the load does not travel forward in a straight path, but rather in a zigzag.

A second problem is that all the rollers in a set must have virtually the same diameter; this forced ancient builders to find whole groups of trees nearly identical in size. Now it may seem that variations in roller diameter shouldn't matter much, because as long as a roller is reasonably round it will certainly roll. The problem, however, is that all of the rollers supporting a given load need to roll forward in tandem, *at the same rate*. If, for instance, one roller has a diameter of 29 cm and another has a diameter of 31cm, the first has a circumference of 91 cm and the second has a circumference of 97 cm ($\pi \times$ diameter, in each case). This means that in each revolution, the bigger roller tries to advance about 7 cm farther than the smaller one. Such differential motion is possible only if there is slipping rather than rolling at one or more of the contact points between a roller and the ground, or a roller and the load. To the extent that such frictional slipping occurs, the advantage of having a roller is nullified, and the effort required to propel the load is increased.

In addition to the above difficulties, a third problem can arise if the load is a giant stone block rather than a cast-iron printing press. Stone is extremely strong when it supports forces of compression, but it fractures easily if it is placed in tension (which in fact is why most of the ancient Greek temples are in ruins today). We take a chance in moving any large piece of stone, because if the supporting forces act to bend the stone even

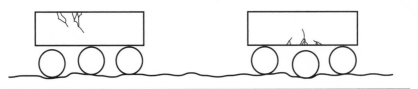

Figure 2.2. A heavy stone can fracture if supported by too few rollers.

a little, it will break at the outside of the bend rather than flexing. In Figure 2.2, we see a stone block supported by three rollers. Because it is impossible for the rollers to be identical, and likewise impossible for the supporting surface to be perfectly flat, situations will always arise where there is a gap between the stone and one or more of the rollers. The great weight of the stone will cause it to sag into such a gap, which promotes a fracture at one of its surfaces. If the support is in the center and the gaps are near the ends, the stone is likely to fracture from the top down; if the gap is in the middle, the stone may break from the bottom up.

So how might an ancient Egyptian engineer avoid fracturing a stone block as it bumped along on an uneven road, supported by irregular rollers? Perhaps by supporting each stone on a very large number of rollers, figuring that the laws of probability would favor enough positive contact to keep the stone from breaking. The more rollers, however, the more cumbersome the moving operation, and the greater the likelihood that some of the rollers would skid rather than roll. A much better method is to first place the stone on a wooden sled, which in turn is supported by a smaller number of rollers. Wood is resilient under both tensile and compressional forces, and this strategy allows a large, heavy stone to be supported fairly uniformly regardless of the inevitable irregularities in how the rollers contact the road. This, in fact, is what the people of Easter Island seem to have done in transporting their giant stone statues (called *moai*), some as tall as 32 feet and weighing up to 89 tons, for distances of several miles over highly irregular terrain.[1]

To Build a Pyramid

Over the centuries, numerous scholars have speculated about how the ancient Egyptians built their colossal pyramids. The largest of these structures, completed around 2680 B.C.E., has a base measuring 230 meters on a side and a peak rising 147 meters above the desert sands. This Great

Pyramid, which took twenty years to build, incorporates thousands of large limestone blocks, each weighing several tons, most of which were transported dozens of miles.[2]

It's an impressive accomplishment to hand-cut seven million tons of stone into small-tonnage blocks, and to transport them all to a building site many miles away, without the help of railroads or tractor trailers. There can be no doubt that this effort was assisted by rollers, and that some of the blocks were transported by boat for part of their journey. The more baffling question, though, is how these blocks were lifted as high as 48 stories.

Clearly, the blocks weren't hoisted vertically, because the sides of pyramids are far from vertical (the angle of the Great Pyramid is 51.9° from the horizontal). Most scholars of the subject speculate that the ancient engineers began by building a ramp that circumscribed the perimeter of the pyramid at a much shallower angle, perhaps 10°, and that laborers dragged the stone blocks up this incline to fill in the pyramid's middle and to extend the ramp to greater heights. Eventually this ramp disappeared into the final shape of the pyramid. Yet this leaves unanswered the question of how laborers might have hauled 20-ton blocks up a surface inclined at about 10°. (The steepest slopes on our modern interstate highways measure only about 3.5°, or 6%.)

Simply dragging the stones up the ramps was out of the question. The friction would be so great that it would take nearly two hundred laborers to pull each 20-ton block. Greasing the surface to reduce the friction was equally impractical, because this would leave behind a slippery mess for the gang of laborers pulling the next block.

So what about rollers, the same mechanism used to transport the stones to the building site? In fact, some archaeologists assume that this is the answer: rollers to get a block to the site, more rollers to get it up the ramp. If we actually try to use a set of rollers on an incline, however, we immediately encounter a new set of difficulties. One issue is the matter of timing in catching each roller as it emerges from behind the load, before it rolls freely down the slope and gathers so much momentum that it can't easily be stopped. A complicating problem is that the rollers released behind the load emerge at irregular intervals, regardless of how regular we are about spacing them when we insert them at the front end of the load. In other words, standing behind a load moving up a ramp, we may need to catch not just one roller at a time, but sometimes two or three at split-second intervals.

Why will rollers pile up this way when they support a load on an incline? Because no roller can maintain contact with both the load above and the surface below unless the spacing between these surfaces is consistently and precisely equal to the diameter of the roller. On a level surface, temporary breaks in contact don't present a problem, because the forward momentum of the roller keeps it rolling. On an incline, however, the roller slows down each time it loses contact with either surface. As a result, the rollers emerge from behind the load at highly irregular intervals, compared to their spacing when they are introduced at the front of the load.

Further, the act of introducing rollers under the front end of a load is not an easy task on an incline. On a level surface, we can simply lay the roller in place well forward of the advancing load, then make a final adjustment so the load encounters the roller head-on. But on an incline, we need to *hold* the roller in place until it contacts the full width of the front edge of the load. If the roller is a little aslant, it can't be straightened out after it is gripped by the weight of the load. A roller misoriented in this manner will send the load careening off to the left or right.

No, the ancient Egyptians would not have hauled huge stones up ramps for so many centuries by using rollers in such a clumsy way. Given that the building stones were cut with square cross sections, it is much more likely that they used the scheme shown in Figure 2.3. Because a square is a regular polygon, it can be transformed into a circle by binding identical circular segments to its sides. In practice, these segments were probably made of wood, and recycled from one block to the next. Wrapping a rope around such an assembly (which is easy because of the clearance beneath) gives an effective means of pulling on the load so it rolls up a ramp. Such a strategy circumvents the complications of using a set of rollers, each of which may have a mind of its own.

This approach is most effective if two parallel ropes are wound around the stone, each rope pulled by its own team of laborers. Two ropes facilitate corrections to the direction of travel; if the block begins to veer to the left, for instance, the team on the left pulls harder while the team on the right relaxes a bit until the direction straightens out. Moreover, with two separate ropes, there will always be one rope to hold the block in place while the other is being rewound.

Figure 2.3. Pulling a building stone up a ramp by transforming it into a roller.

There is yet another important advantage to turning the load into its own roller—a quantitative mechanical advantage. By using the system I've just described, ancient builders may have reduced the effort required to draw a load up an incline to just about *half* the effort required if they'd used a set of rollers beneath the load.

How can this be? One way to view what happens is shown in Figure 2.4. As a circle-encased block rolls through one revolution, it travels up the ramp a distance equal to πD, where D is its diameter. Meanwhile, the rope unwinds an additional distance equal to the perimeter of the block, which is $2D\sqrt{2}$. Thus, the total length of rope that has unwound in one revolution is $\pi D + 2D\sqrt{2}$. Again, this leads us to a constant-ratio situation:

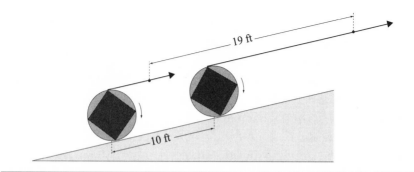

Figure 2.4. A point on the unwinding rope travels about 1.9 times as far as the distance the encased block rolls forward.

$$\frac{\text{effort distance}}{\text{load distance}} = \frac{\pi D + 2D\sqrt{2}}{\pi D} = \frac{\pi + 2\sqrt{2}}{\pi} = 1.900316\ldots$$

The size of the block doesn't affect this ratio. Provided that a square block is encased in circle segments that transform it into a roller, and the rope is wound around the block itself, we end up pulling the rope about 1.9 times as far as the distance the block rolls forward. This is significant, because whenever an effort force moves farther than the load, we gain something in return: a reduced effort requirement. This effect, referred to as a *mechanical advantage*, was well known to the ancient Egyptians (and to every other civilization that moved heavy objects on a regular basis). The mechanical advantage in this case is defined as the ratio of the load's resistance force to the effort force required to move it. This ratio is approximately the same as the ratio of the effort distance to the corresponding load distance:

$$\text{mechanical advantage} = \frac{\text{load resistance force}}{\text{effort force}} \approx \frac{\text{effort distance}}{\text{load distance}}.$$

Let me illustrate with some numbers. In both of the scenarios below, we will move a 20-ton stone block up a 10° ramp. And in both scenarios, the load resistance force will be 1,100 lb, which includes the combined effect of gravity and rolling resistance.

Scenario 1: We pull the block up the ramp on rollers. Because our rope is fastened to the block itself, the effort distance equals the load distance, and the mechanical advantage is 1. This means that the effort force equals the load resistance force, and to push or pull the load up the ramp requires an effort force of 1,100 lb.

Scenario 2: We transform the block into a roller, as in Figures 2.3 and 2.4. Now, as we previously calculated, the effort distance is 1.9 times the load distance, and the mechanical advantage is 1.9. This means that the effort force is lower than the load resistance force by the factor $\frac{1}{1.9}$, so to pull the 20-ton load up the ramp requires an effort force of just 580 lb.

Thus, if we transform the load into its own roller, we can move it up an incline with little more than *half* the effort force, as compared to the effort requirement if we had placed a set of rollers under the load.

From the Roller to the Pulley

Rollers are a great help in moving a heavy load along a horizontal plane. To lift a heavy weight vertically is a much bigger challenge. Levers provide a sizable mechanical advantage, but only over small distances, and other devices like screw jacks and hydraulic lifts require fine machining tolerances. Yet the ancients obviously did lift many heavy objects through great vertical distances. The stones that span the capitals of the columns of the Parthenon, for instance, weigh around 9 tons apiece, and were hoisted vertically a distance of 10 to 11 meters. This and similar tasks have long been accomplished with systems of pulleys.

A pulley is a variation on the roller, supported at its center so it can roll in place as a moving cable maintains contact with a portion of its circumference. A single pulley does nothing more than change the direction of the force in the cable. A *system* of pulleys, however, can do quite a bit more, as shown in Figure 2.5. Here, we see a 200-lb load that in the first case requires a 100-lb effort to lift it, and in the second case is lifted by just a 50-lb effort. The mechanical advantages result from the fact that the load force is divided among a number of supporting segments of the cable: two in

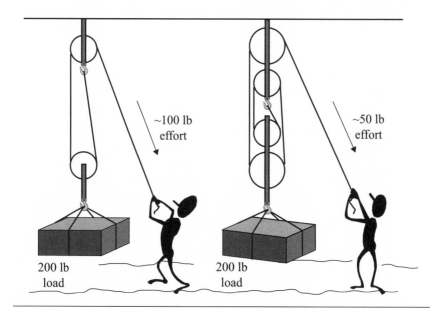

Figure 2.5. A coupled system of fixed and movable pulleys can generate a significant mechanical advantage.

the first case, and four in the second case. If we support a 200-lb load with four parallel cable segments, then each segment supports 50 lb. To lift this load by pulling on just one of these supporting cable segments, we need to pull with only a little more than a 50-lb effort, which is equivalent to a mechanical advantage of about 4. Of course, we don't get this mechanical advantage free of cost; the trade-off is that the effort force needs to travel four times as far as the load. Still, this is usually a good bargain, for it makes it possible to lift heavy objects we wouldn't come close to budging by using human muscle directly.

Wheels, Axles, and Bearings

Wheeled vehicles relieve us of the need to catch and insert rollers beneath a moving load. The ancient Egyptians and Sumerians certainly recognized this principle and its advantages, for they not only had wheeled chariots, but also wheeled utility wagons. Yet one thing the ancients did *not* do was to transport their heavy building stones on wheels.

And why not? Because the underlying principle of the wheel, and the thing that makes it a wheel rather than a roller, is that the load is prevented from moving forward farther than the axle. In contrast, we have seen that a load on a roller travels twice as far as the corresponding forward movement of the roller. An ideal roller does not slip against the load as it rolls, but a wheel's axle, ideal or not, must always slip in the region where it supports its load. Such slippage generates heat and wears away the materials that rub against each other. With the bearing materials available to the ancients, the transportation of heavy loads on wheel-and-axle assemblies would have generated so much friction that frequent mechanical failures would have occurred. Through most of human history, therefore, the use of wheels has been restricted to transporting relatively light loads.

Let's look more closely at how one constrains a wheel to travel at the same speed as the load it supports. A wheel is typically attached to an axle that rotates with it, and the function of this axle is to support the load. The load bears on the axle through a contact region called, appropriately enough, a *bearing*. The earliest, and simplest, bearing was the yoke bearing (Fig. 2.6a), which allowed the axle to rotate yet prevented the load from traveling any farther than the center of the axle. Although this arrangement generated considerable friction in the small contact regions between the

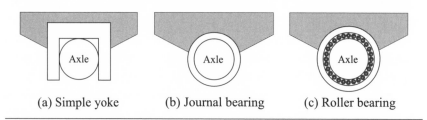

(a) Simple yoke   (b) Journal bearing   (c) Roller bearing

Figure 2.6. Supporting a wheel requires a bearing that rides upon the wheel's axle. The bearing, if not precisely round, is subject to considerable wear and premature failure.

yoke and the axle, for light and slow-moving wagons it did its job reasonably well.

An improvement came with the journal bearing (Fig. 2.6b). Here, the bearing surrounds the axle concentrically. By maximizing the area of contact between axle and bearing, this geometry minimizes localized wearing of the materials. It also provides a means to apply lubrication without having it immediately drip back out. The journal bearing, however, requires that both the axle and the bearing be fairly true circles, and the technology for accomplishing this simply and effectively was not developed until the late Middle Ages. The earliest journal bearings, used on Egyptian and Roman chariots, were not particularly round and therefore could be used to carry only light loads for short distances.

By the late nineteenth century, engineers began developing high-speed engines and other rotating devices that generated destructive frictional heating in even a precision and well-lubricated journal bearing. In mechanisms where friction needed to be minimized as much as possible, ball and roller bearings were adopted (Fig. 2.6c). In these devices, a set of steel balls or cylindrical rollers is constrained to roll in a grooved track (called a race) surrounding the axle; these balls or rollers move at about half the speed of the axle surface. Such bearings must be machined to extreme precision, using tools whose complexity greatly exceeds the complexity of their product.[3] Even ball and roller bearings, however, do not completely eliminate slipping and friction, as is evidenced by squealing when their lubricant runs dry. Inevitable manufacturing errors confound engineers when they seek to apply basic geometrical principles.

For most of human history, wheeled vehicles were uncommon devices, not because the wheel hadn't been conceived, but because it couldn't be put into practice in a reliable way. In fact, among the pre-Columbian

Native American cultures (some of which were fairly advanced technologically), we see wheel-and-axle assemblies only in toys. Europeans made some advances during the Middle Ages, particularly in the bearings that supported large water wheels and windmills, but even in those times windmills often burned to the ground when extreme winds drove the mechanism at too high a speed and its wooden bearings burst into flame. Only in the late eighteenth century, with the onset of the Industrial Revolution, did mechanics develop the machinery needed to make axles and bearings round enough to reliably carry heavy loads. A wheel's circumference can be a little out-of-round and it will still roll along a road fairly well, but its axle and bearings must be very close to true circles if they are to enjoy reasonable longevity.

CHAPTER 3

THE CELESTIAL CLOCK

Tumbleweeds roll across the plains, maple seeds spin as they flutter to the ground, and turbulent streams swirl into whirlpools and eddies. When threatened by floods, some varieties of African ants clump themselves into a floating ball that is kept rolling by the mad scramble of each ant trying to get to the top. A fish snatching a fly from the surface of a still lake sets off a series of circular ripples that expands outward from the point of the disturbance. Pebbles in streams tend to get worn into roughly circular shapes, and even in antiquity what child would not have noticed that such pebbles roll if you toss them with a spin onto a hard surface?

Although these natural examples are inexact geometrically, they do suggest the idea of roundness, and this concept can be extended to applications where artificial roundness may be useful or desirable. Long before written history, humans ornamented their bodies with rings, bracelets, and anklets, and many built dwellings in the shapes of cylinders and hemispheres. A particularly practical early application was the earthenware pot. If we were to take some clay and shape it into a vessel for carrying water or perhaps grain, what shape would we choose? Certainly not an unnatural shape like a cube, where the corners are likely to get bumped off (just as happens to pebbles in a stream). No, it's much more effective, as well as easier, to make a round vessel. A round pot also offers this bonus: it uses the minimum possible amount (and weight) of material to enclose a given volume. If we carry water in a round pot, rather than using a container of some other shape, we are assured that we've maximized the fraction of our effort that goes into carrying the water itself rather than its container. We will return to the geometrical reasons for this efficiency of the sphere later, in chapter 6.

From studies of those Stone Age cultures that survived into recent times—for example, the Native Americans and the tribes of sub-Saharan Africa—we know that the circle emerges over and again in the religious rites and sociopolitical rituals of societies that live in intimate harmony with nature. While on Earth the roundness of fruits, seeds, tepees, and earthenware pots is linked to survival and efficiency, the more nearly perfect circles in the night sky suggested a benevolence and order at the highest levels of the natural environment. Our earliest written records, as well as the archaeological evidence of times preceding, all confirm that early humans looked to the skies in their search for perfection and universal truths.

Wheels in the Heavens

With the exception of the disks of the sun and full moon, circles in the sky are rather subtle. No one notices at first glance that the sun, moon, and stars move in (nearly) circular paths relative to the Earth-based observer; what we see instead is that the sun rises and sets, and that specific stars appear in different parts of the sky at different times. There were, however, many thousands of starry nights in ancient Mesopotamia and Egypt, where shepherds stayed out all night guarding their flocks. Looking above the northern horizon, they saw the ancient counterparts of today's constellations (Cassiopeia, the Big Dipper, and the Little Dipper, for instance) always maintaining consistent shapes and standing in the same relationship to one another. Over the course of each night, this whole assembly of star-patterns seemed to swing around a single fixed point, as if the stars were stuck to a giant spinning wheel. Figure 3.1 shows a time-exposure photograph of today's northern night sky, in which the stars leave obvious circular streaks as they rotate about the North Star during the course of the night.[1] Despite the lack of cameras in ancient times, many observations over many nights would eventually lead the astute observer to a similar mental image of a rotating sky. In this motion, it would seem, lies the "perfect" circle, its center riveted to a single spot above the northern horizon, night after night, season after season, year after year.

True, there are some celestial objects whose motions deviate from this scheme, in particular the visible planets and the occasional comet, and in the briefest of intervals the meteor or "shooting star." Over the course of

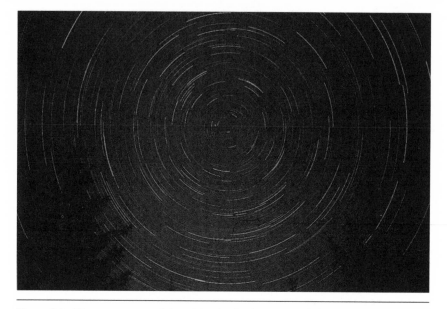

Figure 3.1. Time exposure of the night sky, looking toward the northern horizon.

a year, the sun and the moon have observable noncircularities in their motions. But for the most part, the motions of the thousands of visible stars in the night sky would have appeared to the ancients to follow circular paths. By comparison, any roundness observed on Earth was but a poor imitation of the geometrical perfection found in the heavens.

Of Sixes and Sixties

In ancient Mesopotamia it was common knowledge that a fairly accurate circle could be drawn by looping a piece of twine over a stake in the ground, then using a stick to draw the loop tight while tracing a continuous curve. If all parts of a curved line are the same distance from a single fixed point, that line has no choice but to close upon itself and become a circle.

But what if we have only part of a circle, say an arc as shown in Figure 3.1? How might that portion of a circle be described? In hindsight, some have suggested that the Mesopotamians should have divided the full circle by tens, into say a hundred or a thousand parts. Instead, as we well know, they divided it into 360 parts, each of these divisions designating one circular degree.

To reconstruct the reasons for the choice of 360 requires a bit of conjecture, but nothing too farfetched.[2] We begin by noticing, as did the Mesopotamians, that it's an easy matter to divide a circle into six equal parts, as in Figure 3.2. All we need to do is mark the points where radius-length segments intersect the circumference. It doesn't take 2π radii to get around the circle in this manner, because our line segments do not follow the arc. Only six radii are needed to return to the starting point on the circumference. Whatever the number of degrees we choose for a circle, then, there is a definite advantage to picking a number that is evenly divisible by six, because such a division can easily be done geometrically.

Beyond this, there is also an advantage to choosing a number that is evenly divisible by four, because one-fourth of a circle is a right angle, and even ancient builders provided most of their structures with horizontal floors and vertical walls. So what numbers are evenly divisible by both 4 and 6? There are many, starting with 12, then 24, 36, 48, and so on up to numbers like 348, 360, and 372. It can hardly be an accident that the second number in this series is 24, the number of hours in a day. But given that any of the numbers in this integer series might have been reasonable choices for defining a circular degree, why was 360 chosen?

A circumstantial fact is that 360, of all the numbers in this series, has the most integer divisors. We might imagine a group of authorities sitting at a table (perhaps a round table) to decide this issue. Those who were

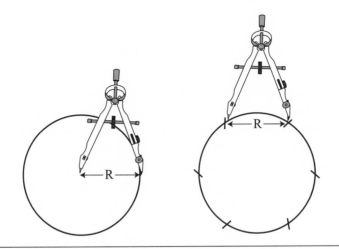

Figure 3.2. Dividing a circle into sixths. The radius of the circle is R.

accustomed to counting on their fingers might have held out for a system based on 10, while no one disagreed that 6 and 4 also needed to be factors because they could be constructed geometrically. Of course, other divisions might also prove useful on occasion. What number can we come up with that is evenly divisible by 2, 3, 4, 5, 6, 8, 9, and 10? The lowest is 360. The fellow arguing for 7 loses, because that would require 2,520 degrees in a circle, which is much too unwieldy a number to impose on society on the unlikely chance that someone might want to divide a circle into sevenths. Besides, 360 happens to be curiously close to the number of days in a year, and this could potentially be a useful relationship (with 360 degrees in a circle, the sun advances through the constellations by about one degree each day). Further, 360 days = 12 × 30 days, and 30 days is close to the phase month of the moon (29.53059 days). As we will see shortly, these latter relations are more than numerological curiosities, for they facilitate the measurement of time and the construction of calendars.

Of course, we have no way of knowing what deliberations actually led to the definition of the 360-degree circle. It's clear, however, that 360 degrees was not a bad choice, or else by now it would have been long abandoned and replaced. No other unit of measurement has come even close to the longevity of the circular degree, currently around six thousand years old.

It was a long time before anyone had a practical need to measure angles to a precision much finer than one degree. Still, the human eye can do considerably better. Standing at the center of a circle, anyone with normal vision will be able to distinguish two points on the circumference if these points are at least $\frac{1}{30}$ of a degree apart.[3] This, it turns out, has nothing to do with the physical size of the circle; the objects can even be distant stars. If the angular separation of a pair of stars is smaller than $\frac{1}{30}$ of a degree, to most people they will merge visually into a single point of light, but if they are separated by more than $\frac{1}{30}$ of a degree, anyone with normal vision, or with vision corrected to normal, will see two separate stars.

This suggests that each degree ought to be divided into at least 30 smaller units. Instead, the number 60 was chosen for reasons of consistency. Begin with a circle, divide it geometrically into 6 sectors, and each of these sectors is 60 degrees. Now divide each degree into 60 "minute parts" of a degree (from the Latin *pars minuta*), or *minutes* for short. This gives 60 minutes in a degree, 60 × 60 minutes in a 60-degree sector, and 60 × 60 × 6 minutes in a full circle.

Later, after the development of optical telescopes in the sixteenth century, it became possible to resolve angles even smaller than a minute. To accommodate this increase in precision, another division of 60 was introduced: a "second minute part" (*pars minuta secunda*), or *second* for short. This gives us 60 seconds in a minute, 60×60 seconds in a degree, $60 \times 60 \times 60$ seconds in a 60-degree sector, and $60 \times 60 \times 60 \times 6$ seconds in a circle. No one has ever had a realistic need to extend this further, say to third minutes, fourth minutes, and so on, because computationally we now find it much simpler to use decimal notation rather than fractions.

By the late nineteenth century virtually all science textbooks had adopted decimal notation, so, for instance, rather than writing 25 minutes as $\frac{25}{60}$ degrees, the more common representation would be 0.41667 degrees. Of course, this is not exact, because the 6's actually repeat to infinity (as we saw in chapter 1). Still, the need to truncate the decimal fraction is no great drawback in science, for no measurement can be exact anyway, and it makes no sense to calculate beyond the precision that can actually be observed and measured.

Yet in certain fields, particularly surveying and navigation, the degree-minute-second convention is still in use today. As an example, and using the standard abbreviations:

$$32 \text{ degrees } 44 \text{ minutes } 27 \text{ seconds} = 32° \ 44' \ 27''$$
$$= 32 + \frac{44}{60} + \frac{27}{3,600} \text{ degrees}$$
$$= 32.74083°.$$

Many hand-held calculators now perform this computation with a single function key, sparing modern surveyors the need to sum these fractions each time they want to calculate such a conversion.

The Sundial

From our observation point on Earth, it appears that the sun rises in the east (at least approximately), crosses the sky, and sets in the west. It is by no means immediately apparent that our Earth itself is rotating toward the east, and that this is the reason for the sun's apparent motion in the opposite direction. So, just like the ancients, let's assume that the sun really does move, and let's explore whether there is any constant ratio that describes this movement.

If we watch a shadow over the course of the day, we find that its length changes in a complicated way. Shadows are long in the morning, grow shorter toward midday, then lengthen again toward sunset, and there is no constant ratio (π or any other number) that relates the length of a shadow to the passage of time. Further, it turns out that all of the great capitals of the ancient world (Athens, Rome, Cairo, Babylon, Beijing, etc.) lay north of the tropics, where the sun never passes directly overhead and therefore the shadow of a vertical object never disappears. Confounding matters even more, the sun does not rise (or set) at the same point on the horizon on any two consecutive days. During the spring, when the days are getting longer, each morning the sun rises a little farther to the north, and after the summer solstice, each consecutive sunrise takes place a little farther to the south.

Despite these complications, it's still clear that the sun's motion is cyclical. This motion can be revealed by orienting a semicircular strip of metal with a pin (a "gnomon") at the center of the arc so it casts a shadow on the strip. Recording the movement of the shadow over time reveals a constant ratio between the sun's *angular* displacement and the lapse of time. Using the Mesopotamian units of angle, and noting that a full 360° circle requires the passage of 24 hours, we have

$$\frac{\text{angular displacement}}{\text{time}} = \frac{360°}{24 \text{ h}} = \frac{15°}{1 \text{ h}}.$$

Indeed, if we track the sun along its path, adjusting for the fact that this path is not perpendicular to Earth's surface and that the actual inclination changes a bit from day to day, we find that the sun always travels through an angle of about 15° each hour. This suggests that a properly oriented protractor can be transformed into a sundial by inscribing an hour mark every 15° (Fig. 3.3).

Of course, an hour remains a fairly crude unit, and a sundial can easily be more precise than this. Divide each hour into 60 small ("minute") parts of an hour, or "minutes," and the same constant ratio tells us that one degree of the sun's motion takes a time of 4 minutes. The Mesopotamian system of sixties makes the ratio of time per degree work out rather nicely to an integer: in 4 minutes of time, the sun travels through an arc of one degree.

The oldest sundial still in existence is an Egyptian one dating from about 1500 B.C.E. The largest sundial ever constructed was built 3,200 years later

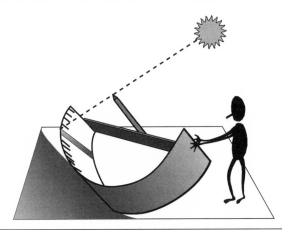

Figure 3.3. Using a protractor as a sundial. The shadow of the gnomon moves through fifteen angular degrees every hour, or one angular degree every four minutes.

in Jaipur, India; it covered almost an acre and had a gnomon more than a hundred feet tall. This device also compensated for the sun's different path on different days of the year. By this late date of 1724, it was no longer necessary to design the device to register the constant ratio of 15° per hour, because trigonometric calculations had made it possible to lay out a non-linear scale on the ground that correlated the shadow of the gnomon with the position of the sun in the sky. This huge Indian instrument was hardly an anachronism, however, because sundials continued to be important in navigation right through the nineteenth century, and as recently as 1900 one highly precise sundial was still being used in France to set the watches of railway men.

Of course, a sundial can't indicate time when it's cloudy, or at night. This limitation spurred the development of other devices that could keep track of time during periods when the sun didn't shine. Some ancient clocks were based on flowing sand, some on water dripping from one vessel to another, and others on the consumption of oil by a lamp. What the sundial continued to provide was a standard for adjusting such timepieces to keep them ticking off hours in pace with the motion of the sun. Place a water clock next to a sundial, mark a line corresponding to the current time, wait an hour and mark another line at the new water level, continue this process, and the result is an instrument that will count hours in step with the sundial, even at times and places where there is no sunlight. Every so often, as necessary, the user can go to a sundial and readjust such a clock.

In fact, in the fourteenth and fifteenth centuries, when it was a matter of civic pride for European towns to boast having a tower clock (and the pendulum principle had not yet been discovered), such clocks were accurate only to around a half-hour per day. It was therefore an important duty of the clock keeper to reset the clock according to the reading of a sundial at least every couple of days.

Well into the twentieth century, the motion of the sun was still the basis for the measurement of time throughout the world; only in 1967 was the standard for timekeeping completely converted from an astronomical basis to the atomic clock. By then, scientists had learned that the microwaves emitted by cesium-133 atoms have a much more stable frequency than the relative motions of the sun and Earth through space. With the development of the atomic clock, the standard for timekeeping collapsed from the astronomical to the submicroscopic. Today, one day is no longer equal to the Mesopotamian $24 \times 60 \times 60$ seconds, or 86,400 seconds. Instead, we now *define* one second to be the time of 9,192,631,770 oscillations of the microwaves in a cesium-133 clock. As a result, the length of a solar day works out to be around 86,400.002 modern seconds, a number that actually varies a bit from year to year.

The small discrepancy between the old solar day and the newer atom-based day accumulates to around one second over the course of a typical year. In fact, while watching a New Year's Eve telecast, we may have heard an announcement that one or two "leap seconds" will be introduced between the old year and the new year. Such "leap seconds" reflect our continuing efforts to keep the current atomic timekeeping system in reasonably close step with the ancient astronomical solar day.

Months and the Calendar

Astronomical motions give us not just one natural time unit, but at least three: the day, the month, and the year. The day, of course, is the time interval from one sunrise until the next. The month originated in the observed cycle of the phases of the moon. The year is based on the cycle of the seasons, and four recurring events: the summer and winter solstices, and the vernal and autumnal equinoxes. But why do we earthlings divide our year into five months of 30 days, six months of 31 days, and a remaining month of sometimes 28 days, sometimes 29? The answer must be sought in records of cultures of the past.

Ancient calendars, particularly those of early Rome, were often abused for political purposes, adding days here and there to lengthen the term of a favored official or to make an upcoming political event fall on a historic anniversary. By the time of the reign of Julius Caesar, the calendar had been modified so much that January had slipped into autumn. In 46 B.C.E., following the advice of his astronomer, Caesar reformed the calendar, added 90 days, and named a new month (and in fact the whole new system) after himself: hence, the month of July and the Julian calendar.

The Julian calendar retained earlier idiosyncrasies like months of unequal length, but it did attempt to synchronize the daily cycle of the sun's motion with the longer annual cycle of the returning seasons. Yet over the next sixteen centuries, it became apparent that the Julian calendar wasn't quite right, because the vernal equinox had slipped forward ten days during this period, from March 21 to March 11. This was a serious problem for many Christians, because the date of Easter, then the most important holy day of the year, was set by tradition as the first Sunday after the first full moon following the vernal equinox. To return Easter to where it belonged by tradition, Pope Gregory XIII decreed in 1582 that the ten days' slippage be added back in, and that the system for determining leap years be modified so that such a discrepancy would never arise again. Despite riots in some communities by peasants who thought that the church authorities had stolen away ten days of their lives, the Gregorian calendar was eventually adopted throughout most of the world. Britain (no great supporter of the papacy) held out until 1752, when in British lands around the world, the day after September 2, 1752 (Julian), became September 14, 1752 (Gregorian). Dates of historical events between 1582 and 1752 are therefore often found to disagree by ten to twelve days, depending on which calendar the writer was using. (George Washington, for instance, was born on February 11, 1732, under the Julian calendar, and although this is what all the contemporary documents say, we see his birthday listed on our current calendars as February 22.) Some nations were even slower to adopt the Gregorian calendar: Japan in 1873, Greece in 1912, and Turkey in 1927, for instance. Although for purposes of global commerce and communication virtually all of the modern world now uses the same calendar, for religious purposes several other calendars still remain in widespread use. We will shortly examine some of these religious calendars in more detail.

Despite its arbitrary length under the current Gregorian calendar, the month traces its origin to a natural cycle of the moon, which goes through a series of phases: new, first quarter, full, last quarter. This cycle is easy to observe, and it would be convenient for timekeeping if it took an integer number of days to complete. Unfortunately, the sun follows one motion, the moon follows another, and the two cycles are not intrinsically synchronized. There is no way to connect the solar year with the lunar month through a purely logical analysis; all one can do is carefully observe many cycles of the moon and compute an average time that reflects nature's actual choice in this matter. This apparently was done fairly accurately by the Arabs as early as the eighth or ninth century, with the result

phase month of the moon = 29.5306 solar days.

In our modern world, we might legitimately question why anyone would care about the number of days in a cycle of lunar phases. But in ancient times, and in some places well into the twentieth century, one would never plan to travel at night unless there would be a nearly full moon to light the road. In fact, some of the founding fathers of the United States belonged to a social group called the "Lunar Society," so-named because it was their practice to meet on the night of the full moon. If their political discussions ran late into the night, their horses would still be able to see well enough to get them back home. Not only is the full moon quite bright, but because it rises at sunset and sets at sunrise, it is the only phase of the moon that provides illumination all night long.

A number of societies have built their calendars on lunar cycles. Among the most interesting is the Muslim calendar, in which each year has exactly twelve months, each month beginning at or very near the new moon. In the days before quartz calendar watches, this allowed Muslims to check the date, at least approximately, just by looking at the moon: if an observer saw a first-quarter phase, it was the seventh or eigth day of the month; if he saw a full moon, it had to be the fifteenth, and so on. Moreover, since the angle between the moon and the sun changes by about 12° each day, with enough practice an early Muslim might even become skilled enough to distinguish the twenty-second of the month from the twenty-third just by looking at the moon.

A Muslim year has either 354 or 355 days, depending on whether or not it is a leap year (11 of every 30 are). Clearly, the Muslim year is too short to keep up with the seasons, and given enough time, any given month eventually falls both in summer and winter. Each 30-year cycle of this calendar contains exactly 10,631 days and 360 months. Taking the ratio of these two numbers, we find that the Muslim month averages 29.53055 . . . days, which differs from the actual phase month of the moon by less than 1 day in 600,000. This is quite an impressive degree of accuracy, considering that year 1 of this calendar corresponds to the Christian year 622 C.E. Since its inception, the Muslim calendar still has not accumulated an error of one full day in tracking the phases of the moon.

In civilizations where people didn't need to travel in the desert at night, seasonal cycles were a more important social concern than lunar-phase cycles. Calendars reflect social priorities as least as much as they reflect astronomical observation and mathematics. The prehistoric ruins at Stonehenge, and at some of the Anasazi sites in the U.S. Southwest, were originally laid out as giant protractors designed to pinpoint the days of the summer and winter solstices (i.e., the longest and shortest days of the year). A solar year, however, does not contain an integer number of solar days; the actual relationship is:

$$1 \text{ solar year} = 365.2421929 \ldots \text{ solar days}$$

There is no evidence to suggest that any future increase in precision measurement will reveal this to be a rational number. As with most measurements, the number of solar days in a solar year seems to be intrinsically irrational, and therefore impossible to divide evenly into smaller units. The Mesopotamians and ancient Egyptians, however, did the best they could and developed calendars of 365 days, which they divided into 12 months of 30 days, with 5 and a fraction days left over. During some periods of history, the 5 leftover days were decreed to be holidays—a bonus from the heavens. Occasionally, as needed, an extra day might be added to keep the seasons synchronized with the calendar; in fact we still do this today with our leap years. There was no easy way, however, to keep *three* independent timekeeping units synchronized: day, month, and year. In most cultures, the decision was made to leave one of these natural timekeeping units by the wayside. For the Muslims, it was the solar year that was sacrificed. In most other cultures, it was the month that was allowed to be adjusted and politicized.

There is, however, one notable exception. The Jewish calendar, which preceded the Muslim calendar historically, was particularly innovative in that it did attempt to keep all three cycles (solar, lunar, and seasonal) synchronized over a long-term average. As in the Muslim calendar, all Jewish-calendar months have either 29 or 30 days, with an average fairly close to the phase month of the moon. Some years have 12 months that total 353, 354, or 355 days (comparable in length to the Muslim year), but there are also leap years of 13 months that total 383, 384, and 385 days. Overall, the average length of the Jewish year is remarkably close to the solar year of 365.2422 days. Obviously, the great thinkers who devised the Jewish calendar had long-term rather than short-term accuracy on their minds.

CHAPTER 4

MATHEMATICS AND THE PHYSICAL WORLD

Pythagoras was born on the Greek island of Samos around 582 B.C.E., and he died in 507 B.C.E. We know nothing about the man's writings, and it's quite possible that he never wrote anything; in ancient times, it was common practice for great thinkers to delegate the documentation of their ideas to their students and followers. We do know that the ideas and teachings of Pythagoras attracted a large following, and that his students developed a religious society that survived for a century or two after his death. These Pythagorean disciples distributed numerous writings, some profound, some apocryphal, and some (unfortunately) ridiculous. Compounding the confusion, only fragments of many of these documents survived through the ages, often after going through numerous transcriptions and translations that may have resulted in various embellishments and omissions. Regardless of the confusion about the details, however, it's clear that Pythagoras did originate a tradition that has had a profound influence on Western thought: the tradition that mathematics can uncover fundamental truths about the physical universe.

Pythagoras was preoccupied with the question of whether numbers and mathematical logic have anything to do with the patterns and order observed in nature. Not only did he observe the natural world around him, but he also tweaked it in various ways to see if it spoke back in a manner that could be interpreted mathematically. In one of his experiments, he took a multiple-stringed musical instrument and strummed a note on one string; then he tuned a second string so it sounded the same musical pitch.[1] Next, he placed a movable bridge under the second string and slid it along the string's length, carefully listening until he located points that were in musical "harmony" with the unbridged string. When he found such points, he compared the length of the bridged string with the unbridged string, and he expressed this relationship as a ratio. The ratios delivered a big surprise—the strings sounded in musical harmony *only* when their lengths were in the ratio of a common fraction:

$$\frac{\text{bridged string length}}{\text{unbridged string length}} = \frac{1}{2}, \frac{1}{3}, \frac{1}{4}, \frac{1}{5}, \frac{1}{6}, \frac{1}{7}, \frac{1}{8} \cdots$$

Now there is nothing strange per se about bridging a string at half its length, one-third, one-fourth, and so on, any more than it would be strange to cut a loaf of bread into thirds or quarters. But unlike the bread, the vibrating string speaks back. And what this string seems to be saying is that it *prefers* to be bridged according to common fractions, because otherwise it produces disharmonious sounds. This finding intrigued Pythagoras; fractions, after all, are but an abstract creation of the human mind. Why should fractions have anything to do with physical musical instruments, unless the products of human mathematics and the workings of the universe are intrinsically intertwined?

The followers of Pythagoras embraced this idea as axiomatic: the essence of the entire universe lies in mathematical patterns. To the extent that they learned to understand numbers and their relationships, then, they would open a door into the mind of Zeus, supreme among gods and possessor of power, rule, and law. To the Pythagoreans, numbers held the answers to all the mysteries of the universe, and the keys to predicting all events that were destined to happen in the future.

This generalization, however, created a problem, in that there were gaps in the ancient numerical system. One could easily count in integers, and this process was easily extended to negative integers. Common fractions were also fairly straightforward, because they were just ratios of integers. All such numbers could be represented by sets or subsets of physical objects (e.g., one-third was simply one element out of a set of three objects). But what about those string-length ratios where the music was out of harmony? If such ratios weren't fractions, what were they?

Pythagoras and his disciples were aware that certain mathematical operations led to numbers that didn't seem to exist. One could easily "square" any number by multiplying it by itself; for instance $3 \times 3 = 9$, and

$(\frac{1}{5}) \times (\frac{1}{5}) = \frac{1}{25}$. But to reverse this process, one had to be careful about what numbers one chose. While it's clear that $\sqrt{16} = 4$, and $\sqrt{(\frac{1}{9})} = \frac{1}{3}$, numbers like $\sqrt{2}$ presented a dilemma. Is there any number which, when multiplied by itself, gives 2 as an answer? And, more important to the Pythagorean philosophy, if we found such a number, would it have any physical meaning? Today we can enter the number 1.414213562 into any calculator, and we can press the key that squares the entry. Try this, and our display will probably show 1.999999999. Most of us might shrug, and say that's pretty much the same as the integer 2. But does it really prove that there is a number such as $\sqrt{2}$? Not at all; it only demonstrates that there is a number that is *close*.

The crowning achievement of the Pythagoreans, and the one they built much of their numerology on, was the proof that there exist noninteger and nonfractional numbers that indeed have a physical interpretation. One such proof runs as follows: In Figure 4.1, we see a square that has been divided into four identical triangles, each having two sides measuring 1 unit in length. The square has sides measuring L units, and the question is to find L. The Pythagoreans noted that the area of the large square is $L \times L$ (or L^2), while the area of each of the four triangles is $(\frac{1}{2}) \times 1 \times 1$, or $\frac{1}{2}$ square unit. The total area of the four triangles is therefore $4 \times \frac{1}{2}$, or 2 square units. Because the total area of the four triangles equals the area of the square, 2 equals L^2. To find L, then, we seek a number that, when multiplied by itself, gives an answer of 2. Yet there is no common fraction

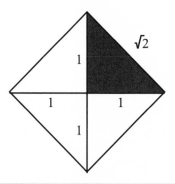

Figure 4.1. The irrational number $\sqrt{2}$ can be interpreted as the diagonal side of a right triangle whose other sides measure 1 unit each. Note that the area of each of the triangles is $(\frac{1}{2})(1)(1)$, or $\frac{1}{2}$, while the area of the large square is 2.

L that meets this requirement; we can square fractions like $\frac{7}{5}$, or $\frac{22}{15}$, or $\frac{707}{500}$, and get close, but we never quite get the integer value 2 as a result. Nevertheless, it's clear from the diagram that $\sqrt{2}$ *does* have a real physical meaning, for it is the length of the diagonal of a right triangle whose sides measure 1 unit each. There is no need to quibble about the fact that a perfect right triangle can't actually be built. The point is this: as a real unit right triangle approaches a perfect unit right triangle, the length of its diagonal approaches a number that *cannot* be expressed as a ratio of integers.

The Pythagoreans generalized this line of logic to apply to any right triangle, even if its sides are unequal, and even if neither side measures 1 unit. The well-known result is that if a right triangle has sides x and y, the diagonal side R must always satisfy

$$R^2 = x^2 + y^2 \text{ (Pythagorean theorem)}.$$

There are many formal proofs of this theorem, some of which are extremely clever, but it is not my purpose to go into them here. Looking instead at the implications, it's clear that in most right triangles we encounter in the physical world, one or more of the quantities x, y, and R must be an irrational number, even in the ideal limit of perfect geometry.

The students of Pythagoras went much further than the famous Pythagorean theorem, and they claimed numerical patterns in nature even when there was no observational evidence that such patterns existed. In their scheme of things, all truths of the universe were reducible to numbers, even such abstractions as justice and beauty. This led to a lot of numerological nonsense, including the casting of people's futures through obscure computations. Ultimately, when they began to interfere with traditional religious customs, the Pythagoreans were driven out of Greece.

Counting

Natural objects don't come with numbers attached. Like circles, numbers are abstractions that exist only in the human mind: even 7 and π are not themselves numbers, but merely *symbols* for numbers. To apply mathematical logic to natural phenomena, we first need to connect our numbers with real physical things. We do this in two ways: by counting and by measuring. These two processes are fundamentally different.

Counting at first blush seems straightforward: we simply place a set of objects in one-to-one correspondence with our set of positive integers. If we count 34 cantaloupes in a wheelbarrow, we expect another observer to get the same answer, and surely no one will get slight variations like 33.91 or 34.28. Using the metric system, everyone still gets 34 cantaloupes. Using Roman numerals, we get XXXIV, which is just an alternative symbol for the number thirty-four.

Yet the consistency of a count by different counters is by no means guaranteed, even in the absence of mistakes. The underlying assumption is that the members of the set be unambiguously identifiable, that is, that all counters know a cantaloupe when they see one. But should a rotting cantaloupe be counted? What about one that is too young and green to eat? Or one that a squirrel has taken a few bites from? Frequently, some counters will legitimately decide differently than others about whether certain objects do or do not belong to the set being counted. The result? An inconsistent count.

Mathematicians can draw Venn diagrams and otherwise define in the abstract which items belongs to a particular set; as a result, counts are never ambiguous in the world of pure mathematics. In the natural world, however, everything is interconnected, and boundaries at best remain fuzzy. Try to count the ducks in a lake, and the issue immediately arises of what constitutes the lake. (Is a mud flat part of the lake? Is an island that frequently floods?) Count the trees in an orchard, and the issue arises of what constitutes a tree. (How big does a sapling need to be to be counted? What about a tree that is nearly dead?) These kinds of issues can sometimes have great practical implications, as when the a nation conducts its census.

Only at the submolecular level does nature sometimes present us with sets of unambiguously discrete objects. Boundaries and definitions of entities are human constructs that nature is under no obligation to respect. We define, and we group, because it helps us describe and understand what we see. But to believe that our definitions and groupings are the reality is to entertain a delusion, and can easily lead us to draw misleading or meaningless conclusions about the world around us.

In grouping and counting, we sacrifice detail to gain a more generalized description. We should therefore not be surprised when two independent counts disagree; this frequently happens (as with death tolls in a disaster), and neither figure need be "wrong." The point to be remembered is this: no numerical count should be accepted at face value without consideration of the empirical contexts in which the number arose. Although we strive for it, certainty remains elusive even in a matter as simple as counting.

Measuring

Suppose that we want to find the height of a certain tree. Here, there is nothing to count, unless it is 1 (that is, *one* tree). Clearly, the process for finding our tree's height will be different from finding a number of cantaloupes. Most likely, we'll choose to use some indirect technique, perhaps based on proportional shadow lengths, or the lumberjack "felling" method of Figure 4.2. This still leaves us another decision to make, and that is our choice of measurement units: it's entirely possible for one person to determine the tree's height to be numerically 46.3, while a second person gets 14.1, and a third gets 556, without any of the three being far from wrong. The point, of course, is that the choice of *measurement unit* has a major effect on the numerical answer:

$$\frac{\text{height of tree}}{\text{standard foot}} = 46.3;$$

$$\frac{\text{height of tree}}{\text{standard meter}} = 14.1;$$

$$\frac{\text{height of tree}}{\text{standard inch}} = 556.$$

Notice that I have described the results as ratios, because that in fact is what a measurement is. Any measured (as opposed to counted) quantity

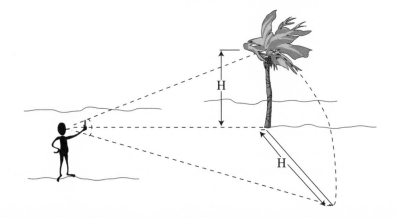

Figure 4.2. The "felling" method for finding the height of a tree. Pivoting a stick held at arm's length establishes the arc the treetop would trace if the tree were to fall. The height *H* is then measured along the ground.

is the result of a *comparison* with a standard unit. Compared to a foot, this tree is 46.3 times as tall; compared to a meter, it is 14.1 times as tall, and so on. In this manner, we could easily have come up with at least eighty different numerical values for the height of the same tree, because there are at least eighty standard linear units we can choose from: yards, leagues, cubits, miles, nautical miles, rods, links and chains, centimeters, and so on. Obviously, then, there is no way a reader can make sense of a naked number without an attached unit of measurement.

Units, however, are not the only difference between measurement and counting. Importantly, we need to recognize that *no measurement is ever exact*. Clearly, the reported height of a tree can't be exact, if for no other reason than the rustling of the topmost leaves in the wind. But what about something more stable: say the distance between two fine scratches in a slab of granite? Can't this kind of quantity be measured exactly? The answer is still no. Every measuring instrument necessarily has a limit to its finest scale division. If we use a ruler divided into 1-millimeter increments, then we can't measure beyond this degree of precision; we might read 206 mm, for instance, but we can't say anything about fractions of a millimeter. Using a more precise instrument doesn't resolve this problem, for even if such an instrument reads 206.347 mm, the digit in the next decimal place still remains unknown. Measurements are never exact, even in principle.

The impossibility of exactness may be viewed this way: When we divide a measuring unit into smaller divisions, we do so through rational fractions. When we measure most physical quantities, however, we find that the digits neither terminate nor repeat; instead, as we increase the precision, we keep getting new digits that do not fall into any pattern. The neutron-proton mass ratio, for instance, has been measured to ten-digit precision, with the currently-accepted result of 1.001378404. . . . All empirical evidence suggests that such physical quantities would be represented fully only by irrational numbers, and that our measured values are but truncated portions of such irrational numbers. Since the rational and irrational numbers belong to mutually exclusive sets, we are forced to the conclusion that no (rational) measurement scale can possibly represent a physical quantity exactly.

There are, of course, a whole range of other practical problems in making physical measurements: imperfections in the instruments, sensitivity to environmental factors like temperature and humidity that may cause an instrument to expand or contract, extraneous variables like voltage surges or vibrations that can throw a reading into error, and even fundamental problems of definition—for example, what does one *mean* by the diameter of the moon? I won't dwell on such matters here, except to point out that even if we could get a handle on all such factors, we still could not make an exact measurement. It's impossible to express physical quantities exactly by using a measurement scale of only rational numbers, and that's that.[2]

When we try to decipher the secrets of nature, we need to be particularly careful about how we draw conclusions from measurements. On the other hand, we can just as easily be led astray by excessive reliance on pure logic that does not seek confirmation in the messy world of measurements. Neither mathematical logic nor empiricism can stand alone if we hope to make progress in revealing the fundamental truths of the physical universe.

The Rise of Formal Geometry

The ancients were well aware that no measurement can be exact. In fact, this was a continuing source of frustration to early merchants, whose profit margins were affected by the accuracy of the scales and weights they used to transact business. It created even greater problems for mapmakers, for there was no way to draw a map without combining measurements that had been reported by many different observers from many different places over extended periods of time.[3] If an explorer measured coastal distances from the deck of a ship, the result never agreed with the corresponding distances measured on the shore itself. If a geographer drew a map that included more than one nation, there was always the question of how to adjust for the different units of distance used in the different countries (which were sometimes changed arbitrarily by royal decree). Today, many of the wild inaccuracies of early maps are apparent at just a brief glance (Figure 4.3). Such geographical distortions reflect the difficulties of measurement more than a lack of skill on the part of the cartographer.

By 300 B.C.E., it was apparent that many geometrical relationships could be described exactly in the abstract, provided one *pretended* that physical constructions and their measurements were exact. The fruits of such geometrical reasoning would then be at least approximately true in the real world. The sum of the angles of a plane triangle, for instance, equals exactly 180° in the ideal case, which suggests that it ought to be close to 180° for real triangles on surfaces that are approximately flat.

46

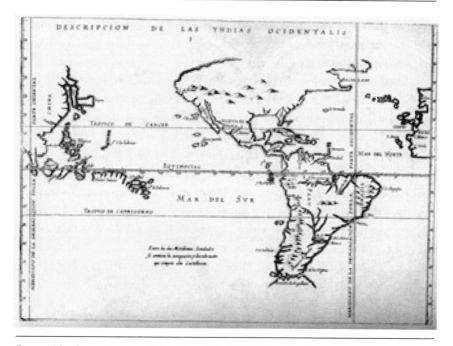

Figure 4.3. An early map of the New World (1622). The distortions are due not to a lack of skill on the part of the mapmaker, but rather to the inherent uncertainties in the available data.

The Greek mathematician Euclid (330–275 B.C.E.), working in Alexandria, devoted most of his career to compiling all the mathematical truths that were then known to the Hellenic world. From the outset he noticed that many of these principles seemed to be based on each other, in a confusing mesh of circular arguments. For instance, are the sides of a right triangle related through the Pythagorean theorem, or does this relationship *determine* a triangle to be a right triangle? Euclid recognized that such issues were far from trivial, for it was possible that a set of mathematical premises might support each other, yet all be contradicted when combined with a more extended body of mathematics. Euclid set himself to sorting out this epistemological hodgepodge so that no tangled arguments would remain. The resulting logical structure started with a set of nine axioms— definitions of concepts like "point," "line," "plane," and "equality." Next were five postulates, which he assumed to be true but would leave unproved (e.g., all right angles are equal to one another). From these postulates, Euclid then laboriously proved every mathematical relationship that was then known. The entire work, *Elements*, filled thirteen books, the first six treating what we now call plane geometry, and the remainder dealing with the theory of numbers and solid geometry. Each of the thousands of conclusions in this extensive work is internally consistent with all the others, and, in turn, with the fourteen fundamental postulates and axioms.

Does Euclidean geometry apply to reality? It seems so, at least to the extent that the Euclidean postulates apply to the physical universe. On the other hand, we can't be sure that these postulates apply with infinite precision, because the only way to explore such a proposition is to measure, and measurement, as we've noted, is inexact. The logical structure of Euclidean geometry did, however, accomplish this: it raised fundamental questions for empirical investigators to examine rather than take for granted. There is no geometrical necessity, for instance, that sunlight travel in straight lines, or that planetary orbits be circles. (In fact, both of these conjectures are wrong, although usually not by much.) We use Euclidean geometry as an abstract model—an ideal standard—against which to compare the real workings of nature. And frequently, Euclidean geometry informs us as much about what our universe is *not*, as about what it is.

The Propagation of Uncertainty

In seeking to make sense of the physical world, we encounter a whole series of dilemmas. Not only are our observations inexact, but the laws of mathematics themselves may not apply exactly in the realm of the physically observable (e.g., there is no such thing as a true circle in the physical world). Still, we needn't throw up our hands in despair and abandon mathematical reasoning. Suppose, for instance, that a wheel's diameter is 28.5 inches give or take 0.5 inch. The half-inch is the "uncertainty" in the diameter, and it may reflect the fact that we've measured slightly different diameters from top to bottom versus side to side, or it may reflect the fact that this is the limit of precision of our measuring instrument. Either way, the uncertainty in diameter will lead to an uncertainty in circumference. Does it make sense to ask *how much* uncertainty?

It does, as long as we keep in mind that we are not dealing with exact numbers in any case. One way to find the uncertainty in this wheel's circumference is to do three calculations:

upper limit on $C = \pi \times$ (upper limit on D) $= \pi \times 29.0$ in. $= 91.1$ in;

lower limit on $C = \pi \times$ (lower limit on D) $= \pi \times 28.0$ in. $= 88.0$ in;

"best value" for $C = \pi \times$ (best value for D) $= \pi \times 28.5$ in. $= 89.5$ in.

In examining the answers, we see that the circumference can be stated as 89.5 in., give or take about 1.6 in. This ±1.6 inches is therefore the *uncertainty* in circumference.

Now, if the wheel in this example were a perfect Euclidean circle whose diameter was exactly 28.5 inches, and it rolled through 1,000 revolutions, it would travel forward a distance of 89,535.3906 . . . inches, or 7,461.28255 . . . feet. Because the wheel cannot be perfect, however, it makes no sense to retain all of these digits. An uncertainty of 1.6 inches in circumference, adding up (in the worst case) over the course of 1,000 revolutions, leads to a total uncertainty of 1,600 inches or 133 feet in the total distance traveled. The uncertainty in diameter, in other words, propagates through our calculation and reveals itself as an uncertainty in the total distance the wheel rolls.

If we quote this particular computational result as 7,461.282552 feet, we mislead anyone who reads it, for we can be quite sure that if we actually rolled the wheel, we would never measure this exact number. A better approach is to state

$$\text{distance} = 7{,}461 \text{ ft} \pm 133 \text{ ft.}$$

Unfortunately, this notation can get tiresome if we do it with every computational result. Instead, as a matter of style and expediency, it's more common to simply round off the answer, so it doesn't misrepresent the precision too terribly. In this example, we might report the result as simply 7,500 feet. The thoughtful reader will take this to mean an answer accurate to within 50 or 100 feet.

People often balk at rounding off numbers this way, particularly when our modern calculators give us all those wonderful digits. The thing to remember is that those digits are based on a purely mathematical logic, while the rounded-off result is probably intended to describe something quite real. In linking the mathematical ideal with messy reality, we always make a big leap. A purely mathematical calculation can never make our wheel rounder, our initial measurements more precise, or the physical universe different from what it is.

Transferring a number from the world of mathematical abstraction into the real world, or vice versa, always involves a judgment call. Uncertainties in measurement can never be eliminated, least of all through pure calculation. The numbers we encounter in our lives are always fuzzy around their edges, and to make valid inferences from such numbers, we need to acknowledge this fuzziness.

So it is in the world of science, where a hypothesis can be rejected on the weight of numerical data, but is never proved absolutely on the basis of such data. So it also is in the world of engineering, where "margins of safety" are routinely figured in to compensate for unknown but expected errors in mathematical prediction. And so it is in the business and finance community, where decisions are always based on imperfect and incomplete numerical information. Mathematics enlightens us about our worlds, but by itself it does not and cannot define the realities of those worlds.

CHAPTER 5

CHARTING
THE PLANET

It is unlikely that any seafaring civilization ever believed that Earth was flat. As we stand on shore and watch a sailing ship disappear into the distance, it does so by degrees: first the hull, then the cabin, then the sails and mast. Conversely, as a sailing ship approaches, the mast and sails become visible while the lower parts of the ship are still hidden by the horizon. Such observations are fairly compelling in their suggestion that the surface of the sea is curved.

Around 500 B.C.E., the Pythagoreans taught that Earth was not just curved, but spherical. They seem to have arrived at this idea by studying the phases of the moon, where the edge of the shadow between the sunlit portion and the dark portion is a semicircular curve. We see a similar curved shadow when holding a spherical ball in the sunlight. By analogy, the Pythagoreans figured that the moon was a sphere, and by extending this to a second analogy, they presumed that Earth was also a sphere.

Although this is not an impeccable argument, it was strengthened considerably after the Athenian philosopher Anaxagoras (c. 500–428 B.C.E.) deduced the cause of lunar eclipses. It had long been known that lunar eclipses occur only when the moon is in its full phase, and it took only a little geometrical intuition to figure that a full moon occurs when the sun and the moon are on approximately opposite sides of Earth. Anaxagoras went further, recognizing that when the moon moved through a region that is almost precisely opposite the sun, Earth's shadow would block some of the sunlight that would otherwise reach the moon, and the event would become an eclipse. Anaxagoras also observed that as an eclipse began, the

Figure 5.1. The curved shadows of lunar eclipses suggested to the ancients that Earth is probably spherical.

edge of Earth's shadow, cast upon the lunar surface, was a semicircular arc (Figure 5.1). This happens regardless of whether the eclipse occurs near moonrise, moonset, or the middle of the night. For Earth to cast a consistently circular shadow, Anaxagoras reasoned, Earth itself must be spherical in shape. By the third century B.C.E., few thinkers disputed that our Earth is round, and many also agreed that Earth rotates on its axis. Some, such as Aristarchus (c. 310–c. 230 B.C.E.) went even further, claiming correctly that the moon is only a fraction of the size of Earth, the sun is many times larger and farther away, and Earth orbits the sun while the moon orbits Earth. Although Aristarchus was essentially right, he did not succeed in convincing many people of his third conclusion—that Earth moves. It would be nearly 1,800 years before Renaissance scientists like Galileo and Kepler would establish beyond any reasonable doubt that our planet Earth orbits the sun, rotates on its axis, and wobbles a bit as it does so.

The essential contribution of the ancient Greeks, however, was not so much in the details of what they thought to be true, but rather in their use of reasoning by analogy to link mathematical truths to physical truths. The logic leading to their conclusion that the world is round may be summarized thus: because little round objects are seen to cast circular shadows, a big circular shadow suggests the presence of a big round object. Although such logic is by no means infallible, it does provide a means for humans to expand their mental vision far beyond the scale of the directly observable. Most of our modern scientific understanding of physical reality rests on a foundation of reasoning by analogy, from the properties of the smallest subatomic particles to the largest galaxies, from the peculiar behaviors of objects traveling near the speed of light to the structure of objects billions of light-years distant.

The Size of Earth

It's a big leap, however, to go from the question of the *shape* of Earth to the question of the *size* of Earth. The ancients of the Middle East had knowledge of less than one-quarter of the world's geography running east and west, and less than one-sixth running north and south. The globe would not be circumnavigated until Magellan's historic voyage of 1519–1522 C.E. No human would set foot even close to the North Pole until Robert Peary's expedition of 1909, and the South Pole would remain unseen by human eyes until 1911.

Yet, around 200 B.C.E., the Greek scientist Eratosthenes determined the circumference of Earth to an impressive degree of accuracy. Obviously, he couldn't do this by actually traveling around the whole of Earth. Instead, his result was based on mathematical logic, the geometrical properties of circles, analogy, and a few key pieces of observational information.

Eratosthenes was affiliated with the great library in Alexandria, Egypt, which was at that time a kind of research institute as well as a repository for most of the world's knowledge. Apparently a traveler had stopped at the library and related his observation that at Syene (near today's Aswan), far to the south, there was one day each year when at midday the sun's rays shined directly down the shafts of deep wells, and tall columns cast no shadows. That day was the summer solstice, June 22 by our modern calendar. Eratosthenes found this tale very curious, because at Alexandria there was no day of the year when the sun tracked directly overhead. Why would the sun do so at some other location? Eratosthenes drew some geometrical diagrams and decided there were only two possible explanations. One, perhaps the sun is very close to Earth, so like a campfire it casts shadows radially in all directions. Alternatively, maybe the sun is very far away, as indeed it appears to be, but the shadows fall in different directions because of the curvature of Earth.

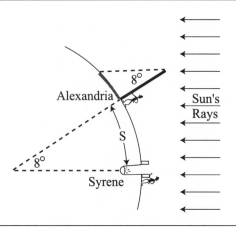

Figure 5.2. Eratosthenes' method for determining the size of Earth. The distance s between Alexandria and Syene was about 4,800 Greek stadia. On this basis, Earth's circumference is around 216,000 stadia, or close to 25,000 miles.

To Eratosthenes, given what he knew about eclipses, the latter explanation was the only reasonable one. And this triggered a stroke of creative genius: a method for finding the size of the world. The geometry is shown in Figure 5.2. If, on the day of the summer solstice, one measures the direction of the sun's rays as they strike Alexandria, this angle θ can be used in a constant-ratio calculation:

$$\frac{\theta}{360°} = \frac{\text{distance between Alexandria and Syene}}{\text{circumference of Earth}}.$$

Eratosthenes used a tall, straight stick and its shadow to measure the angle θ, getting a result of about 8°. The only other information he needed was the surface distance between Alexandria and Syene. Although Eratosthenes already knew this value approximately, he apparently hired someone to travel this distance and make a better measurement. The result, which was quite an accomplishment for the times, was that Syene lay 4,800 Greek stadia south of Alexandria.

Now 8° is $\frac{1}{45}$ of 360°, so a constant-ratio calculation suggests that Earth's circumference should be 45 × 4,800 stadia, or about 216,000 stadia. Although today we no longer measure distances in stadia, we do know that a stadium was, by definition, 600 Greek feet, which corresponds approximately to 606 feet 9 inches in the U.S. system. On this basis, it appears that Eratosthenes established the circumference of Earth at about 130 million

U.S. feet, or a little under 25,000 miles (about 40,000 km). This is consistent with our modern value to within a few percent: an astounding success for extending small-scale geometry, by analogy, to find the size of a planetary sphere much too large to measure directly.

The Equator and the Tropics

Even before Eratosthenes, it was a well-known fact that the sun rises directly in the east and sets due west on only two days of the year: the equinoxes, roughly September 23 and March 21 on our modern calendar. During spring and summer (in the Northern Hemisphere), the sun rises and sets north of an east-west line, while during fall and winter it rises and sets to the south. The northernmost sunrise occurs on the day of the summer solstice (about June 21), and the southernmost sunrise is the winter solstice (about December 21). The paths of the sun on the solstices and equinoxes are shown in Figure 5.3. At latitudes north of Syene, there is no time of the year when the sun passes directly overhead.

Eratosthenes was well acquainted with this annual variation in the sun's path across the sky. He further noticed, in making angular measurements, that at noon at the summer solstice, the sun is 47° higher in the sky than at noon at the winter solstice. On the dates of the equinoxes, this angle is split in half, 23.5° (actually closer to 23.46°). This led him to picture the globe as being encircled by three imaginary circles, as shown in Figure 5.4. The central circle, the equator, marks the set of points where the sun passes

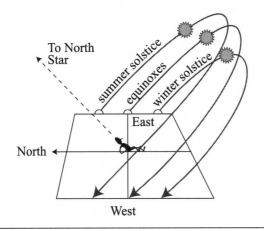

Figure 5.3. The paths of the sun on the solstices and equinoxes.

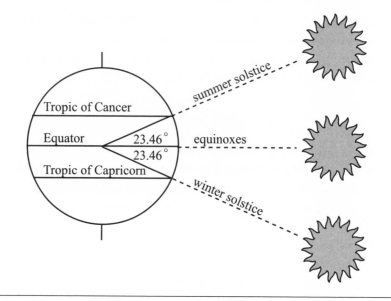

Figure 5.4. Definitions of the equator and the Tropics of Cancer and Capricorn.

directly overhead on the dates of the equinoxes. The other two circles are the Tropic of Cancer and the Tropic of Capricorn, which are the sets of points where the sun passes directly overhead at noon on the solstices. Of course, this places ancient Syene on the Tropic of Cancer. Eratosthenes defined these circles without ever traveling to these latitudes himself. But many centuries later, when Middle Eastern and European explorers did begin venturing into Earth's equatorial region, they had a way of knowing where they were, compliments of Eratosthenes.

Latitude

The site of ancient Syene now lies beneath the waters of Lake Nasser. Even when Syene existed, however, it made more sense to describe the location of Alexandria as 31.5° north of the equator, rather than as 8° north of Syene. The equator is a natural reference circle for describing the north-south positions of all points on Earth, because angles from the equatorial plane can be measured on any clear night of the year.

At Alexandria, the North Star (Polaris) hovers 31.5° above the northern horizon, and to the observer on Earth, it stays in this position throughout the night as the other stars rotate about it (as we've already seen in Fig. 3.1). As the seasons change and the sun travels in a different path, the North Star still hovers at the same point, 31.5° above the horizon in Alexandria. It is no accident that 31.5° is also the angle between Alexandria and the equator. This angle is referred to as the *latitude* of Alexandria.

Figure 5.5 shows that any observer's latitude is equal to the angle of elevation of the North Star above the horizon. At Syene, this angle was 23.5°, in modern Moscow it is 56°, and at points on the equator it is 0° (i.e., the North Star is right on the horizon). If we stood at the North Pole, we'd always see the North Star directly overhead, 90° above the horizon. Measure the angular elevation of the North Star from any point in the Northern Hemisphere, and we immediately know our latitude. South of the equator this is a bit more difficult, because there doesn't happen to be a prominent star aligned with the southern polar axis. Even so, we can determine our southern latitude approximately by sighting on the center of a star pattern known as the Southern Cross, and we can improve our accuracy by averaging a series of such bearings.

This principle of sighting on the North Star was used by navigators to locate their north-south positions even in ancient times. By 150 C.E., Claudius Ptolemy's world atlas included twenty-seven maps with clearly labeled lines of latitude, or "parallels," so any traveler could correlate his own North Star measurements with a map. By the time of the great voyages of discovery, no competent navigator would ever suffer much confusion

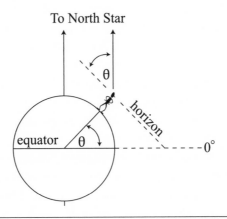

Figure 5.5. The angular elevation of the North Star above the horizon equals the latitude of the observer.

about where he was as regards north and south. East and west, however, presented a much bigger challenge.

Longitude

To locate a point on any surface requires two separate measurements, and the surface of Earth is no exception. In addition to the north-south latitude, a second number, the longitude, is needed to establish one's position in an east-west direction. Yet there is no east star or west star to measure from, because as we look east or west all of the stars appear to move. And because there are no natural poles pointing east and west, there is no "natural" reference circle for drawing a set of parallels running north and south.

Instead, since ancient times, the north-south lines of longitude (or meridians) have been drawn to converge at Earth's poles rather than running parallel to one another. The location of the 0° line of longitude, the prime meridian, is arbitrary; Claudius Ptolemy ran it through the Fortunate Islands off the northwest coast of Africa, which was the westernmost point he knew about. Other mapmakers ran their prime meridians through the Azores, the Cape Verde Islands, Rome, Copenhagen, St. Petersburg, Philadelphia, and a host of other places. Today, it runs through Greenwich, England, just east of London.[1]

The coordinate system we lay over the world's surface, then, is akin to a giant tangerine. Each wedge-shaped piece of the tangerine is narrower at the top and bottom and fatter in the middle, yet the wedges all fit together to make a complete spherical tangerine. Similarly, the lines of longitude converge at the poles and have their greatest separation at the equator. Imagine that we're looking at a peeled tangerine from a point directly above the stem, and we get a picture like Figure 5.6, which shows the convergence of the lines of longitude at the North Pole. The prime meridian is the 0° line, and we measure longitude east (positive) or west (negative) from this reference, up to 180° in either direction.

But how do we make this measurement? Before the days of global-positioning satellites, there was only one method, and that was based on local time. Because Earth rotates through 360° in 24 hours, we've seen that each 15° of rotation takes an hour and each degree of rotation takes 4 minutes. Suppose that we use a sundial to measure our local time on the deck of a ship, and that we carry another clock that tells the local time at the

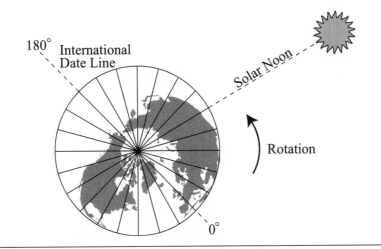

Figure 5.6. Lines of longitude as they appear from a vantage point above the North Pole.

prime meridian. Then, if the two clocks differ by 4 minutes, we are 1° from the prime meridian, and if they differ by 8 hours, we are 8 × 15°, or 120° from the prime meridian. If our local time is earlier than Greenwich time, we are at 120° west longitude, and if our local time is later, we are at 120° east.

Although this concept is simple enough, its practical application was horribly difficult, because before the eighteenth century, no one knew how to build a clock that could be set at the home port and then would continue to keep accurate time on a voyage lasting many months, on a pitching and rolling ship, while subject to great variations in temperature. So, although the concept of longitude traces its origin to the ancient Greeks, in fact it did not yield to accurate measurement for many centuries, until the invention of the marine chronometer by John Harrison in 1759.[2]

Time Zones

When sundials were used to establish local time, the time of day varied continuously as one traveled east or west. Suppose that we set a watch by a sundial, so it reads 12:00 noon when the sun is at its highest point in the sky. Then we travel 1° of longitude to the west (a surface distance of 69 land miles or less), and we check our watch against another sundial at noon. What we find is that our watch reads 4 minutes early, even if all of the instruments are accurate. Travel another 1° of longitude to the west, and the discrepancy between our watch and a local sundial grows to 8 minutes. Travel 15°, and our watch is off by a full hour.

Until the nineteenth century, this continuous variation of local time created no serious difficulties, because no one traveled very fast and there were no methods for long-distance communication that would cause anyone to have to deal with time discrepancies between different locations. With the advent of train travel and telegraph, however, it became confusing for each town to maintain its own local timekeeping system. To remedy this confusion, the world was divided into twenty-four time zones, each averaging 15° in longitudinal width. Towns within a given time zone agreed (usually, at least) to keep their clocks synchronized, regardless of their slight variations in solar time. Travelers crossing from one time zone to the next would now set their clocks backward (going west) or forward (going east) by precisely one hour. This was a major simplification over dealing with continuous variations in local time.

Geometrically, it would be an elegant thing to cover the globe with identical time zones, whose borders coincided exactly with lines of longitude 15° apart. In practice, however, such a purely geometrical system would result in many time-zone boundaries cutting through towns and cities— a situation where the clocks on one side of a street would have to be set an hour different from the clocks on the other side. There are obvious reasons for avoiding situations like this, and in fact if we look at a map of the world, we immediately see that the boundaries between time zones jog around many major population centers. Only on an average are time zones 15° in longitudinal width.

Although each individual country retains the authority to decide what timekeeping system it will use, to facilitate trade and communications most nations have chosen to adopt time zones that are consistent with those of the contiguous countries. In the United States, which spans five time zones, the last changes were set forth in the Uniform Time Act of 1966, which also provided for the use of daylight saving time, but which permitted individual states (particularly those that were split into two time zones) to legislate their own deviations. In Indiana, for instance, some counties follow eastern time, some follow central time, and some flip back and forth between eastern and central time during different months of the year.

There is one additional problem a practical time-zone system must address. Suppose that we fly completely around the world, dutifully turning our wristwatches back one hour each time we enter a new time zone to the west. When we get back to our original airport, our watches will read correctly, because we've turned them back twenty-four times. Unfortunately, our calendar will read one day early. To avoid this difficulty, there needs to be one time-zone boundary where a *full day* is added if we cross it going west, and a full day is subtracted if we cross it going east. This boundary is called the international date line.

Now if the international date line were to slice through any populated region, it would generate unbelievable confusion for the locals: driving to a bank, for instance, and arriving a day earlier or later. Clearly, the date line ought to be situated where it affects no one but international travelers. And, by luck, it turns out that the line of longitude directly opposite the prime meridian does cut through a portion of the western Pacific Ocean where there are very few islands or land masses. The longitude here is 180° west, and it is also 180° east. The international date line has been drawn so it jogs about this 180° line of longitude and misses every island. Only in Antarctica and on the floating ice mass around the North Pole can one actually walk across the international date line.

The international time-zone system is a civil system rather than a physical one. It is no longer connected very closely with the motion of the sun, and in most locations the sun will not be highest in the sky at noon. Further, if we are near the eastern portion of a time zone, the sun will rise around an hour earlier than if we are near the western edge of the same time zone. Such is the consequence of adapting the natural clock to serve human needs.

The Nautical Mile

Maps were drawn long before the invention of a practical marine chronometer in 1759, and although early maps weren't very accurate by today's standards, many did indicate longitude as well as latitude. Clearly, the early discoverers weren't totally ignorant of their transient east-west locations, and their longitudinal estimates, albeit imprecise, were still considerably better than wild guesses. But how did they do this?

The process, once again, began with the recognition of a constant ratio. Because the world is 25,000 miles in circumference at the equator, the ratio 25,000 miles ÷ 360° gives us 69.4 miles of surface distance per degree, or 1.15 miles of surface distance per arc-minute. So, if we can estimate the surface distance we've traveled, we can convert this to the angular units of longitude and latitude.

On oceans, the surface distance was figured to be a ship's forward speed multiplied by the time it maintained this speed. One early practice for finding a ship's speed was to toss a log or other floating object into the water off the bow, then clock the time until the ship's stern passed this object. Dividing the length of the ship by the time measurement gave the speed. Clearly, there are many sources of potential error in such an approach, and I scarcely need to elaborate on them. Even so, this clumsy procedure was much better than guessing one's longitude. If, for instance, a ship began the day near the equator at a longitude of 29° 25′ W, and it averaged 8.3 miles per hour to the east for the next 10 hours, the surface distance traveled would be 10 × 8.3 mi, or 83 mi, and the corresponding longitudinal change would be 83 ÷ 1.15, or 72 arc-minutes. The navigator could then compute the new longitudinal position as 29° 25′ minus 72′, or about 28° 13′ W.

In the days before calculators, the need to repeatedly divide surface distances by 1.15 to get arc-minutes was a nuisance at best. To streamline this process, as well as to avoid possible errors in long division, a new measurement unit was adopted: the nautical mile, equal to 1.15 land miles. One nautical mile north or south corresponds to a change in latitude of one arc-minute. Similarly, one nautical mile east or west is a change in longitude of one minute, provided that we are near the equator. At more northerly or southerly latitudes, a correction can be introduced to compensate for the fact that the lines of longitude converge toward the poles.

A speed of one nautical mile per hour was then called a "knot." If, for instance, a ship maintained a speed due west of 12 knots for 4 hours, the navigator could easily figure that they had traveled 12 × 4 = 48 nautical miles, which corresponds to 48 arc-minutes of longitude, provided they were near the equator. If the latitude was 40° rather than 0°, a correction factor of 1.305 would be applied (this is 1/cos 40°), so here the longitudinal change would be 1.305 × 48′, or about 63′. For changes in latitude, whose lines run parallel, no correction factor is needed. A latitudinal change of one arc-minute north or south always corresponds to a surface distance of one nautical mile.

Today the nautical mile is used for air navigation as well as for ocean navigation, because relative to the size of Earth, aircraft don't fly very high above the surface (seldom more than about $\frac{1}{10}$ of 1% of Earth's radius).

Navigating over long distances, and drawing large-scale maps, however, always require us to take Earth's spherical geometry into account. The nautical mile has succeeded as a measurement unit because it streamlines the connection between surface distances and the angular coordinates that locate our positions on our spherical planet.

The Arctic and Antarctic Circles

Shown on every globe and drawn (in projection) on every world map, the Arctic Circle connects points around Earth at 66.54° north latitude, and the Antarctic Circle is at 66.54° south latitude. But why would anyone attach special significance to such an unusual number? Clearly, it must not be the number itself that is important, but rather some geographical concept that gives rise to this numerical value. In fact, the latitude of the Arctic Circle is but an artifact of a geometrical relationship, and it is this underlying geometry that is significant.

We've already seen that the Tropic of Capricorn is a circle of latitude where the sun is directly overhead at noon around the date of December 22, and that this happens at a latitude 23.46° south of the equator. On that day of the year, the sun's path is quite low in the sky in the Northern Hemisphere, and the farther north we go, the lower the path of the sun. At points 90° north of the Tropic of Capricorn, we reach a circle of latitude where the sun barely peeks over the horizon at noon at the winter solstice. This is the Arctic Circle. North of this circle, the sun does not rise at all during the first day of winter. The Arctic Circle has a latitude of 66.54° N only because the Tropic of Capricorn happens to lie at 23.46° south latitude, and Earth's spherical geometry requires that there be a 90° angle between the two.

Meanwhile, looking at Figure 5.7, we find a circle of latitude around the South Pole that basks in sunlight all twenty-four hours of the day when locations within the Arctic Circle are in twenty-four-hour darkness. Six months later, of course, the roles of the Arctic and Antarctic Circles are reversed; then, on the first day of summer, points within the Arctic Circle experience continuous sunlight, and points within the Antarctic Circle experience continuous darkness. The fact that these phenomena occur at latitudes of 66.54° north and south is directly related to the ancient observation that at the summer solstice, the noonday sunlight lit the bottom of deep wells in Syene (latitude 23.46° N). Although travelers and explorers

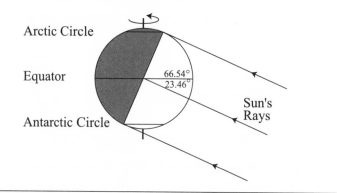

Figure 5.7. The sun's illumination of Earth on the winter solstice results in a northern region of twenty-four-hour darkness and a southern region of twenty-four-hour daylight. These regions are bounded by the Arctic and Antarctic Circles.

observe the sun's motion across the Arctic sky to be quite different from what it is in the Tropics, geometrically these widely spaced observations are intimately linked with each other. The connection is that Earth is approximately spherical. The geometry of ideal spheres indeed applies fairly well to real (although approximate) spheres, even if they are the size of a planet.

Banding Earth

Although our planet appears to be a sphere from a distance, up close it has an obvious surface texture. Is there any way that Earth's mountains and valleys might be a necessary consequence of its approximately spherical geometry? Let me propose a geometrical problem that may initially sound artificial. After we solve this problem, I'll discuss what this simple mathematics suggests about the real tectonic processes that have wrinkled the crust of our planet.

The problem is this: Assume Earth is a perfectly smooth sphere with no mountains or valleys, and suppose that we run a steel band around the equator, draw it up tight, then overlap the ends and fasten them together. Now imagine that Earth's temperature rises a fraction of a degree and this 25,000-mile band expands by 20 feet, making it a bit loose. If we rearrange this slightly loose band so it is everywhere an equal height above the earth's surface, what is that height?

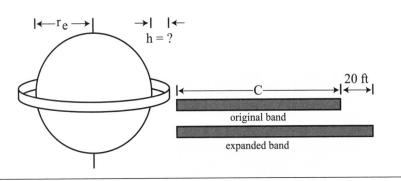

Figure 5.8. Banding Earth. If a band stretched snugly around Earth were to expand in length by 20 feet, what would be its new height above the surface?

The geometry is shown in Figure 5.8. We can approach this problem in a variety of ways, but it's simplified a bit if we state the relationships in symbolic sentences. If C is the circumference of the expanded band, and r_e is the radius of Earth, then

$$C = 2\pi r_e + 20,$$

and

$$C = 2\pi(r_e + h).$$

The first relationship says that the band's circumfence is 20 feet longer than the circumference of Earth itself. The second says that the band's circumference is 2π times its new radius. Because both statements are valid, we can equate the two expressions,

$$2\pi(r_e + h) = 2\pi r_e + 20,$$

which is equivalent to

$$r_e + h = r_e + (^{20}\!/_{2\pi}),$$

or

$$h = {}^{20}\!/_{2\pi} = {}^{10}\!/_{\pi} = 3.18 \text{ (ft)}.$$

In other words, with an expansion of just 20 feet over Earth's entire circumference, the band ends up higher than a kitchen countertop!

Notice that this answer has nothing to do with the actual size of our planet, for nowhere did we need to put in a numerical value for Earth's radius (r_e) or its circumference (C). The numerical answer is equally valid for any sphere: a planet, or a basketball, or even a spider egg. Take any sphere, run a band around its equator, add some arbitrary length L, rearrange the expanded band so it is centered around the original sphere, and the band has no choice but to end up at a height of $L \div 2\pi$ above the surface. Most people find this a surprising result, even after they've worked it out for themselves.

But has a band ever actually been run around Earth? Yes, indeed. The forces of nature did this about 4.5 billion years ago, when Earth's crust first solidified. Today our planet remains mostly molten rock, overlaid with a thin crust only some 4 to 40 miles thick, or an average of less than 1% of the radius of the planet. In effect, the crust may be viewed as a set of circumferential bands that encase Earth's molten interior. Suppose that one of these outer bands of solid rock increases in temperature by a tiny amount, say 0.015 °F, and as a result it expands by about 20 feet, the number I used in the computation above. This expansion drives the crust upward; not just a little, but (as we've seen in the calculation) by several feet.

This leads us to the following picture: Heat emanating from Earth's interior changes the temperature of the crust unevenly, and regions that are warmed expand while places that cool off contract. The crust is thrust upward in some places, while in other places it is drawn apart. The result is a spectrum of geophysical phenomena: earthquakes, volcanoes, and the growth of mountain ranges (Mt. Everest, for instance, grew about 26 feet between 1850 and 1950, and in the Alaskan earthquake of 1964, the surface in some locations was thrust upward by as much as 30 feet). The temperature fluctuations that drive this mechanism are very small and can't easily be measured, for we are talking about temperature changes *within* the crustal bands, at depths of several miles. The conclusion is nevertheless inescapable that significant geophysical surface phenomena result from tiny irregular changes in the planet's circumference. The algebraic calculation does not predict where such events will occur, but it does give us a mathematical analogy to explain them.

And this brings us "full circle" to the fundamental assumption of my problem, for I started by saying that we'll *assume* Earth is a perfectly smooth sphere with no mountains or valleys. Given this assumption, and the practical reality that temperatures within the crust cannot remain perfectly constant, it follows that surface features like mountains and valleys will develop quite naturally. These features will be tiny compared to the size of the planet, but on the scale of human-sized geometry, they will be significant. Paradoxically, it is the geometry of ideal circles and spheres that provides our insight into why a perfectly smooth Earth is a physical impossibility.

CHAPTER 6

SURFACE AND SPACE

One memorable morning during my junior year in high school, and in the fashion of those times, several hundred of us were given a test to measure our IQs. The third question on the test read something like this: "A farmer wants to know the area of a field. He can find this area by (a) adding the length and the width, (b) multiplying 2 times the width, (c) multiplying the length times the width, (d) dividing the length by the width, or (e) adding twice the width to twice the length."

I read the question, thought about it awhile, then left it blank and went on. But it kept bothering me, and a few minutes later I went back and read the question again. The more I thought about it, the more confused I got, for nowhere did the question describe the shape of the field. Growing up in southwestern Pennsylvania, the one thing I could be sure of was that the field would *not* be a rectangle. Farm boundaries followed streams, crests of hills, and roads that snaked along the contours of the land. A field with four sides was a rarity, four straight sides were rarer still, and two pairs of parallel sides meeting at right angles were so unlikely that the situation wasn't even worth considering. None of the answers supplied were valid in the general case. I finally decided that the test's author hadn't thought things out very well, and I quickly filled in the rest of my answer sheet, then dozed off until the papers were collected. When the results came in, my guidance counselor was aghast that my IQ had dropped twenty points in two years.

Today, however, I'm a bit more sympathetic to the person who wrote that test, despite the Midwest bias. The question wasn't really about farm fields. It was about area and dimensionality—attempting to assess whether the test taker conceptualized the difference between surface and linear measure. Had I not been quite so provincial in my teenage thinking at that time, stuck on the fact that the farms I'd seen weren't rectangular, I might have recognized that option (c) was the only dimensionally correct answer and the only reasonable choice.

Surface Measure

Surfaces are by definition two-dimensional, which is to say that it takes two linear measurements to locate each point on a surface. Connecting a set of such points defines a shape that encloses a surface *area*. Clearly, a given area may be enclosed by a multitude of different shapes. In agricultural applications, where one is less interested in the specific shape of a field than in the amount of agricultural product it will support and the amount of labor it will require, various nondimensional measures of land area were adopted in centuries past. In Germany, where grain was at one time measured in a capacity unit called the *Scheffel* (equal to about 50 liters), a Scheffel of land was the area that could be sown with a scheffel of seed. The obsolete English *daieswork* (day's work) was about 2,700 square feet, and in France the area of a vineyard was specified in units of *ouvrée hommée* (man's hours), equal to about 180 square feet. In like manner, the British *acre* was originally equal to an average day's plowing; only later was it standardized as the equivalent of 66 furlongs (each 660 feet in length) with an average spacing of one foot, for a total of 43,560 square feet.

For most applications, however, it is more convenient to express areas in units of a reference square. If a square measures 1 foot on a side, its area is by definition one square foot (1 ft^2), and if it measures 1 meter on a side, its area is one square meter (1 m^2). Notice that although there are 3.281 feet in one meter, there are 10.76 square feet in one square meter. This, of course, comes about because a square meter measures 3.281 ft by 3.281 ft, and the product of these two numbers is 10.76 ft^2. Although this idea is simple enough, it can lead to misunderstandings if we communicate a quantity of, say, 100 square yards of carpet but forget to mention the "square" part. One hundred *square yards* of carpet is equivalent to a length of 20 yards cut from a roll 5 yards in width, which is quite different from 100 yards of the same carpet.

Appendix A lists the formulas for calculating the areas of common shapes. Each of these formulas must involve a multiplication of two linear dimensions — for example, a base times a height, a length times a width, or a radius times itself — because that is the only way a square unit can arise in the result. Conversely, a formula that does *not* result in a square unit cannot possibly be a valid recipe for calculating an area. For instance, it's easy to remember that $A \neq \pi D$, because a diameter D has a linear unit of measure, whereas an area must have a square unit. Similarly, we can be sure the area of even a complicated and unfamiliar shape would not be found by $L^2 W \sqrt{H}$, because again such an expression cannot possibly result in a square unit.

The area formula for any polygon can be established by dissecting that polygon and rearranging the bits until we get a recognizable simpler shape. A parallelogram, for instance, can be transformed into a rectangle by slicing a triangle off one end and tucking it up against the other end; this establishes that the area of a parallelogram is the same as the area of the equivalent rectangle: its base dimension multiplied by the perpendicular height. But can we apply this logic by transforming a *circle* into a rectangular shape?

It turns out we can, provided we are willing to go through a method of successive approximations that gets closer and closer to transforming a circle into a rectangle. In Figure 6.1, we see a circle that has been divided

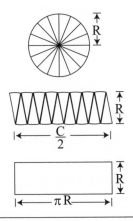

Figure 6.1. Finding the area of a circle by dividing it into sectors. As we increase the number of sectors, we find that their rearrangement more closely approximates a rectangle of length πr and height r. The area of this rectangle, and that of the original circle, is therefore πr^2.

into sixteen identical sectors. In the second diagram, we've pulled these sectors apart and rearranged them so they point alternately up and down. The resulting shape is a crude rectangle, with a height equal to the radius of the original circle and a base roughly equal to half the circumference of the original circle. This establishes that the area of the circle is approximately $A = (r)(\frac{C}{2})$. Since $C = \pi D = 2\pi r$, this is equivalent to $A = (r)(2\pi\frac{R}{2})$, which is equivalent to the familiar formula $A = \pi r^2$. With only sixteen sectors, of course, this remains an approximate result. But what would happen if we divided the circle into a hundred sectors, or a million, or a billion? Would the result change? Clearly, we don't actually need to do this cutting, because we can see that as the circle is sliced into a larger number of sectors, the rearranged sectors will become an increasingly improved approximation to a perfect rectangle. Our conclusion, then, is that the area of a circle is indeed equivalent to the area of a rectangle whose base is $\frac{C}{2}$ and whose height is r, or, in other words, $A = \pi r^2$.

This raises the following question, which was first explored in ancient times: Is it possible to construct a square whose area equals the area of a given circle? By "construct," the Greeks meant that only straight edges and compasses would be used. We can loosen this requirement a bit and allow measurements with rulers that have been marked off in the usual integer increments of rational fractions. So, if we are given a circle, can we draw a square that has the same area?

The answer is that we can come arbitrarily close, but the equality of the two areas will always remain approximate. This would be so even if our ruler were perfectly accurate (which of course it can't be). To reach this conclusion, we merely need to notice that we seek the length L of a square whose area L^2 is the same as the area πr^2 of a given circle. Equating these two areas, $L^2 = \pi r^2$, leads us to the requirement that $L = r\sqrt{\pi}$. So, if $\sqrt{\pi}$ were a rational number, it would be possible to transfer a radius from a circle and use this to lay out a square of corresponding area. Pi, however, is a transcendental number, and so is its square root. No circle, therefore, can be squared except as an approximation.

What does it mean, then, to describe a circular area as, say, 25 square centimeters? The words themselves say that this circle has an area equivalent to that of 25 unit squares each 1 cm on a side, or to a single larger square measuring 5 cm on a side. Yet, without crunching numbers, there is no direct way to connect the diameter of this circle, 0.70898154 . . . cm, to the 5-cm side of the equivalent square. Although we usually find it convenient to describe circular areas in square measurement units, this does not imply that circles can actually be squared geometrically.

Square measurement units are not, however, necessary. Beginning in the early twentieth century, the fledgling electrical industry began to specify the circular cross-sectional area of electrical wires in a unit that had no direct relation to squares.[1] This unit, still in use in the United States, is the *circular mil*, usually abbreviated *cmil*. One circular mil is defined as the area of a circle whose diameter is one mil (one one-thousandth of an inch). Extrapolating from this definition, it follows that a circle with a diameter of 3 mils has an area of 9 cmil, and a 6-mil-diameter circle has an area of 36 cmil. Using these units, the formula for the area of a circle is simplified to $A = D^2$, and a factor of $\pi/4$ has been eliminated. This formula, however, muddles the units, for a circular mil is not the square of a linear mil, but rather

$$1 \text{ cmil} = \pi/4 \text{ mil}^2.$$

Fortunately, the use of the circular mil as an area unit has not caught on for other applications. Square units prevail not only because they are much easier to visualize, but also because they preserve unit consistency in calculating areas, regardless of the actual geometrical shape. It would be considerably more confusing to apply the theorems of geometry to the physical world if we adopted a different unit definition for each different shape: square meters, circular meters, elliptical meters, pentagonal meters, and so on.

The Scaling of Areas

Scientists inquire into the workings of nature by examining individual cases, then trying to *induce* the general governing principles. Mathematicians take the reverse approach, beginning with abstract axioms and postulates, then *deducing* the specific cases (which may or may not have anything to say about physical reality). Scientific generalizations sometimes call for a mathematical system that had never been thought of before (e.g., Isaac Newton's invention of calculus to describe the governing principles of orbital mechanics). Conversely, pure mathematics sometimes predicts physical effects no scientist had ever thought to look for in nature (e.g., solitons, fractals, and black holes). The inductive scientific process and the deductive mathematical process clearly complement one another in enabling us to decipher the fundamental structure of our world.

Let's look at the specific question of the scaling of areas, and speculate on how someone might approach it inductively. The question may be posed as follows: Is there a ratio that tells us how the area of a circle increases as its diameter is increased?

Because our hypothetical investigator knows that $A = \pi r^2$ for a circle, she begins by trying to get a feel for the problem by doing a few specific calculations. She considers three circles, with diameters of 1.00 inch, 2.00 inches, and 3.00 inches, and computes the three corresponding areas:

$$A_1 = \pi r_1^2 = \pi(0.50 \text{ in})^2 = 0.785 \text{ in}^2,$$

$$A_2 = \pi r_2^2 = \pi(1.00 \text{ in})^2 = 3.142 \text{ in}^2,$$

$$A_3 = \pi r_3^2 = \pi(1.50 \text{ in})^2 = 7.069 \text{ in}^2.$$

These numerical results reveal that doubling the diameter increases the area by a factor of $3.142 \div 0.785$, or 4.00, and tripling the diameter increases the area by a factor of $7.069 \div 0.785$, or 9.00. But are these results generalizable, or do they relate only to the particular numbers we've used here?

The power of manipulating numbers symbolically rather than arithmetically is that this frequently reveals deeper patterns. Our investigator starts again, but now she generalizes slightly by writing that her three circles have radii of r_1, $r_2 = 2r_1$, and $r_3 = 3r_1$. She then computes the following ratios of the areas:

$$\frac{A_2}{A_1} = \frac{(\pi r_2^2)}{(\pi r_1^2)} = \left(\frac{r_2}{r_1}\right)^2 = 2^2 = 4.$$

and

$$\frac{A_3}{A_1} = \frac{(\pi r_3^2)}{(\pi r_1^2)} = \left(\frac{r_3}{r_1}\right)^2 = 3^2 = 9.$$

Although these results are the same as her earlier arithmetical conclusions, what she finds here is that the ratio of areas has nothing to do with the absolute sizes of the circles. Instead, the ratio of areas equals the square of the *ratio* of radii (which is the same as the square of the ratio of diameters). If, as a special case, one circle has triple the diameter of another, then the larger one has nine times the area, regardless of whether the first circle is the size of a thumbprint or the size of a galaxy.

Our investigator seeks to generalize further. If the ratio of the radii of two circles is some arbitrary factor K (rather than the integers 2 or 3), she notices that the ratio of the two areas becomes

$$\frac{A_2}{A_1} = K^2, \quad \text{where } \frac{r_2}{r_1} = K.$$

Thus, if a circle's radius is scaled by a factor of 1.5, its area increases by a factor of 1.5 × 1.5, or 2.25, and if the radius increases by a factor of 10, the area increases by a factor of 10 × 10, or 100. Clearly this rapid growth of area arises because area is a two-dimensional measure, while radii (and diameters) are one-dimensional. If a circle's diameter is increased by a factor of 10, for instance, this not only makes the circle 10 times as wide but also 10 times as high. The overall effect on the area is a factor of 10 × 10, or 100.

As a final generalization, our inductive investigator notices that this result is not specific to circles, but that it applies to any other shape as well. Regardless of the shape, the area is always some constant multiplied by the product of two linear dimensions. If each of these two dimensions is scaled by some arbitrary factor K while preserving the shape, clearly the area is scaled by a factor of $K \times K$, or K^2. The shape can be a swimming pool in the shape of a guitar, for instance, and if we contract for a second guitar-shaped pool 1.73 times as long and 1.73 times as wide, we can be guaranteed that the swimming surface *area* has increased by a factor of 1.73^2, or about 3. Here, our larger pool has about 3 times as much surface to swim on as the smaller pool, even though the linear dimensions have been scaled up by a factor of only 1.73.

A mathematician would approach this problem in reverse order, beginning with the fundamental definitions of one- and two-dimensional measure, establishing that area measure *always* scales by the square of the linear factor, then deducing that the circle is but one of an infinity of shapes to which this theorem applies. It is, however, only because both the mathematician and our scientific (inductive) investigator both arrive at the same conclusions that we can be assured the result is a valid description of area scaling in the *physical* world. Standing alone, neither deductive (mathematical) nor inductive (scientific) reasoning can affirm fundamental truths about physical reality.

Are there practical implications of this result on the scaling of areas? Most definitely. Figure 6.2, for instance, shows the standard diameters of water pipes used in the United States. Clearly, large-diameter pipes have greater carrying capacities than smaller pipes, but how much more? If we picture the water flowing in a pipe, it's clear that the flow fills the whole cross section, not just the diameter. As a result, the flow rate (in gallons per minute, for instance) is proportional to the pipe's cross-sectional area. Compare two pipes, and if one has 1.5 times the diameter of another, then the larger pipe's area is $(1.5)^2$ times the smaller pipe's area, and the larger pipe carries $(1.5)^2$ or 2.25 times the flow, all other factors being equal. Using the same logic, if an 8-inch pipe is to be replaced by a bundle of half-inch pipes running side by side, it takes $(8/0.5)^2$ or 256 of the smaller pipes to carry the same flow as the single larger one.

Another instance where area scaling is important is in cables and supporting columns; the strengths of such structural members are proportional to their cross-sectional areas. If we double the diameter of a steel cable, all other factors being equal, the larger cable will be four times as strong, and if we triple the diameter, the larger cable will be nine times as strong.

This scaling effect also works in reverse, and I'm aware of at least one industrial accident that happened as a result. A rigger in a steel mill reported to his foreman that a cable on a crane was damaged, and he insisted that production be halted until the cable could be replaced. The foreman inspected the 1-inch steel cable, agreed that it was unsafe, and directed the rigger to replace it. The rigger explained that there wasn't enough 1-inch cable left in the warehouse; all they had was ½-inch cable. At this, the foreman directed the rigger to use the half-inch cable and to *double it up*. The result: a snapped cable, a 10-ton coil of steel falling 25 feet, damage to machinery, and a nearly tragic outcome for several workers in the area. What happened, of course, is that the foreman erroneously assumed that the strength of a cable is proportional to its diameter, when in fact its strength is proportional to its cross-sectional area. To equal the strength of a 1-inch cable takes *four*, not two, ½-inch cables bundled together. Two ½-inch cables side by side are actually only half as strong as a single 1-inch cable.

We don't need to be in a steel mill to find area scaling useful; all we need to do is order a pizza. Which is the better bargain: a 16-inch pizza for $13.00 or a 12-inch pizza for $9.00? If we work out the price per inch

Diameter	Relative Area
0.5 inch	1.0
0.75	2.25
1.0	4.0
1.5	9.0
3.0	16.0
4.0	64.0
6.0	144.0
8.0	256.0

Figure 6.2. Standard pipe sizes in the United States. Each doubling of the diameter gives four times the flow capacity.

($.81 versus $.75) and conclude that the 12-inch pizza is cheaper, our logic is all wrong. Why? Because we don't eat the pizza along its diameter. In terms of the actual amount of pizza, the 16-inch pizza is the better deal, because

$$\left(\frac{16}{12}\right)^2 = 1.78 \quad \text{(ratio of areas)},$$

while

$$\frac{13.00}{10.00} = 1.3 \quad \text{(ratio of prices)}.$$

In other words, using these numbers, the 16-inch pizza allows us to eat 78% more pizza for a paltry 30% in added cost.

Measuring Space

Space is three-dimensional, which is to say that it takes three independent linear measurements to locate a point in space. Sometimes a set of spatial coordinates defines a continuous closed surface: a milk carton, a cereal box, a fuel tank, or a hot-air balloon. To describe the amount of enclosed space, we can draw upon a wide variety of capacity units: quarts, liters, gallons, pints, teaspoons, fluid ounces, cords, barrels, and so on. Most of these units trace their origin back many centuries, and their current definitions can be found in most dictionaries.

The fundamental way to measure enclosed space is to express it as a *volume* rather than a capacity. The reference unit for volume is the space enclosed by a cube measuring one unit on a side: the cubic foot (ft^3), cubic meter (m^3), cubic yard (yd^3), and so on. Because $1 \ yd^3 = 3 \ ft \times 3 \ ft \times 3 \ ft$, or $27 \ ft^3$, and similarly, $1 \ m^3 = 3.281 \ ft \times 3.281 \ ft \times 3.281 \ ft = 35.31 \ ft^3$, it's important to use the modifier *cubic* when describing a volume unit. Omitting this adjective implies a linear measure and a linear rather than cubic relationship between units.

Appendix B lists some formulas for calculating the volumes of common geometrical shapes. It's noteworthy that each of these formulas involves the product of three linear dimensions. Moreover, if we look for constant ratios between volume and linear dimension, the formulas tell us that we are on a wrong track. For a sphere, for instance, the ratio V/r or V/D is not constant, but rather grows dramatically as the diameter increases. On the other hand, the ratio V/r^3 *is* constant (equal to $4\pi/3$), regardless of the size of the sphere. This constant ratio is equivalent to the statement that volume is proportional to the cube of the radius.

The Scaling of Volumes

Suppose that we have three spheres, with diameters of 2.000 ft, 4.000 ft, and 6.000 ft. How do their volumes compare? This is similar to the question we examined for circular areas, and once again our inductive investigator can answer by just doing the arithmetic:

$$V_1 = (\tfrac{4}{3})(\pi)(1.000 \text{ ft})^3 = 4.189 \text{ ft}^3,$$

$$V_2 = (\tfrac{4}{3})(\pi)(2.000 \text{ ft})^3 = 33.51 \text{ ft}^3,$$

$$V_3 = (\tfrac{4}{3})(\pi)(3.000 \text{ ft})^3 = 113.1 \text{ ft}^3.$$

On this basis, we can conclude that $V_2/V_1 = 8.000$, and $V_3/V_1 = 27.00$. Clearly, these numbers show that the volume of a sphere increases rapidly as the radius is increased by a relatively modest amount. Specifically, doubling the radius (or diameter) increases the volume eight times, and tripling the radius results in twenty-seven times the volume.

Notice, however, that once again we can reach a more general conclusion algebraically. If two spheres have radii that are related by a linear scaling factor K so that

$$r_2 = Kr_1, \quad \text{and} \quad D_2 = KD_1,$$

then the ratio of volumes is

$$\frac{V_2}{V_1} = \frac{[(\tfrac{4}{3})\pi r_2^3]}{[(\tfrac{4}{3})\pi r_1^3]} = \left(\frac{r_2}{r_1}\right)^3 = \left(\frac{Kr_1}{r_1}\right)^3 = K^3.$$

If, for instance, one sphere has five times the diameter of another, we conclude that the volume of the larger one is 5^3, or 125 times the smaller volume. Similarly, if one sphere has ten times the diameter of another, the volume of the larger one is 1,000 times the smaller volume.

Where does all this extra volume come from? From the fact that a sphere is three-dimensional. In increasing the diameter tenfold, we are making the sphere ten times as wide, ten times as high, and ten times as deep. The result is an increase in volume by a factor of $10 \times 10 \times 10$, or 1,000.

As for practical implications, they abound. An object's weight (or more technically, its mass) is proportional to its volume, so if we have a solid iron ball 1.00 inch in diameter with a mass of 33.1 grams, we can expect that a second iron ball 5.00 inches in diameter will have a mass 125 times as great, or about 4,140 grams. If these balls are being shot from guns, it would be a grand mistake to expect that shooting the 5-inch ball should take just five times as much propellant as the 1-inch ball. In fact, it will take about 125 times as much propellant for the larger ball, because 125 times the mass is being shot from the gun.

The real power of dimensional thinking, however, is that it is not restricted to spheres. If *any* solid shape is scaled up or down by the same factor K in all three linear dimensions, this preserves the shape but alters the volume by the factor K^3. Suppose we look at the classified ads and find a used 22-ft sailboat priced at $8,000 and a used 30-ft sailboat priced at $16,000. We'd prefer the larger boat, but we wonder if it's a good deal. On the basis of length, it doesn't seem to be, because the larger boat is only 1.36 times as long but it's priced twice as high. Yet, it turns out that (assuming comparable condition and quality) the larger boat is probably the better value. Why? Because the larger boat is not just 1.36 times as long, but also about 1.36 times as wide and around 1.36 times as deep. Its total volume is $(1.36)^3$ or 2.52 times the volume of the smaller boat. Thus the larger boat contains roughly 2.5 times the amount of material, and probably took about 2.5 times the amount of labor to produce. Since prices reflect labor and materials (rather than length), the 30-ft boat should be priced at about $20,160 to be comparable in value to the 22-ft boat at $8,000. Using this logic, if the 22-ft boat is a bargain at $8,000, then the 30-ft boat at $16,000 may be an even better bargain.

Strength and Weight

In 1726, the English author Jonathan Swift published *Gulliver's Travels*, in which his hapless hero is shipwrecked and has a series of adventures on four fictitious islands. Given that Swift was more interested in political satire than in geometry, it's hardly fair to criticize his geometrical reasoning too harshly. One of Gulliver's adventures, however, does provide an interesting example of fallacious dimensional scaling.

Gulliver's second landfall is the island of Brobdingnag, a land populated by giants with all the features and proportions of normal humans, but who stand around 60 feet tall. They are a peaceful and generous people (unlike the tiny Lilliputians on the first island, who are belligerent far out of proportion to their size). Unbeknownst to the writer, however, a human 60 feet tall has a serious flaw: he can't walk, and he can't even stand. In fact, he'd probably have a hard time crawling.

To arrive at this conclusion, let's assume that Gulliver is 6 feet tall, weighs 200 lb, and is strong enough to carry a total weight of 400 lb. A Brobdingnagian, ten times as tall, has 10^3 or a 1,000 times Gulliver's volume, and therefore weighs about a thousand times as much as Gulliver, or 200,000 lb. Meanwhile, the Brobdingnagian's leg bones and muscles have 10^2 or 100 times the cross-sectional area of Gulliver's corresponding anatomical features. As we have seen, the strength of any cable or supporting column does not depend on its height, but rather on its cross-sectional area. So, if the giants' muscles and bones are made of the same stuff as normal human muscle and bone, the giants must be a hundred times as strong as Gulliver. Given that Gulliver's legs can support 400 lb, a Brobdingnagian should therefore be capable of carrying around 40,000 lb.

The result of this logic is that the giants weigh some 200,000 lb, while their legs can support only around 40,000 lb. Clearly, such a creature could not walk upright. In fact, the worst threat Jonathan Swift's Brobdingnagians could pose is that one might roll over on you.

Given, however, that Brobdingnagians are fictitious, does this story really have any relevance? Yes; it turns out that the same scaling effect can be found throughout the biological world. Every once in a while, for instance, a school of whales will become disoriented and swim onto a beach as the tide goes out, stranding them high and dry. When this happens, most of them die. But why would whales die out of the water, when they're mammals that breathe air just as we do? The reason is that a whale quickly wears itself out trying to expand its lungs against its great body weight unless it remains in a state of neutral buoyancy. The only way the forces of evolution ever managed to create a mammal as big as a whale was to adapt it to life in the sea. In fact, the skeleton of a whale suggests that its ancestors were land animals, for a whale's flippers contain articulated finger bones, and its horizontal tail has standard mammalian foot bones. Presumably, as protowhales grew larger and larger, they spent more and more time in the water, which in turn favored their growing even larger, until today modern whales can no longer survive on land.

Move in the other direction, to smaller animals, and we find that they tend to become very strong in relation to their weight. Who among us, for instance, hasn't become exasperated trying to walk a small dog that has a mind of its own? As the size of an animal is reduced, its weight decreases much more rapidly than its strength. Squirrels can perform incredible acrobatic maneuvers, and frogs can easily jump fifteen times their body length. Yet even with a running start, the best Olympic athlete can't jump as far as five times his height. The analogy begins to weaken when we skip from one phylum to another, but it's certainly worth mentioning that insects — ants in particular — can often be seen carrying objects many times their own weight over great distances.

But let's leave biological organisms, which present a host of fuzzy variables that can affect strength and weight, and look briefly at physical structures that are more easily quantifiable. One assignment commonly given in introductory physics classes in high schools and colleges runs something like this: Build a bridge out of wood and glue that has a free span of at least 40 cm, weighs less than 70 grams, and can support 10 kilograms. The numbers may vary, but using the ones given here, the student is asked to build a structure that can carry about 143 times its own weight. Although students initially may be intimidated by this challenge, meeting the requirement actually turns out to be easy, and the only way to fail is to ignore some fundamental laws of physics. I personally have had students build bridges that weighed less than 70 grams yet successfully supported as much as 60 kilograms (132 lb), or nearly 900 times their own weight. In one case, after we used all the weights I had available and the bridge still didn't fail, the student set his bridge spanning two bricks on the floor, then *stood* on it. It failed only when he lost his balance and accidentally twisted it sideways.

But these were models. Can such designs be scaled up to real bridges? Suppose we took a successful bridge model that had a 200:1 strength-to-weight ratio and scaled it up, using the same proportions, the same materials, and the same design. Is there any limit to the distance it could span? Absolutely. If we increased the span by a factor of 200, this particular bridge would need all its strength just to support its own weight. Anything longer, and it would fail under its own weight. Why? Because if it were 200 times longer, and all its structural members were proportionally larger, the bridge would weigh 200^3 times as much but would be only 200^2 times as strong. In scaling up this bridge model by a linear factor of 200, then, we've *reduced* its strength-to-weight ratio by a factor of 200, and this particular bridge design is now at its size limit.

While there are a variety of other arithmetical ways to arrive at this same conclusion, there is an inherent power to thinking in terms of ratios. One of the advantages of ratios is that they easily lend themselves to verbal reasoning.

Surface-to-Volume Ratios

Suppose that we want to build a container that holds a specified volume, say 1,000 cubic feet. To choose among the many shapes we could use, we pose the following question: What shape will enclose the required volume within the least surface area? In Table 6.1, we see the results of some calculations. Here, we consider six different shapes, each of which encloses an identical volume of 1,000 cubic feet. The dimensions of these shapes are given, along with their calculated surface areas. Note that although the volumes are the same, the surface areas vary considerably.

Now the point is this: if we are building a container to hold a volume of 1,000 ft³, and if we are interested in using the least amount of material

Table 6.1. The surface area needed to enclose a constant volume of 1,000 cubic feet, using a selection of different shapes. A sphere encloses the volume within the least surface area

Shape	Dimensions (ft)	Surface area (ft²)	Volume (ft³)
Rectangular solid	L = 7.071 W = 7.071 H = 20.00	666	1,000
Cube	L = 10.00	600	1,000
Circular cylinder	R = 6.830 H = 6.830	586	1,000
Hemisphere	R = 7.816	576	1,000
Circular cylinder	R = 5.420 H = 10.84	554	1,000
Sphere	R = 6.204	484	1,000

to enclose this volume, clearly the best shape we can choose is a sphere. Next to this, the most efficient shape is a cylinder whose radius is half its height (which is the same as having a diameter equal to the height). A cube requires even more surface area to enclose the same volume, and a non-cubical rectangular solid requires yet more. We can conclude that if we are paying for material based on its surface area, and we are using the material to construct containers, the most economical shape is the sphere. If we decide to avoid spheres for other practical reasons (e.g., they tend to roll), the next-best shape is a cylinder whose diameter equals its height. This result has long been known to manufacturers of canned goods.

Besides the cost benefit, a low surface-to-volume ratio may offer other advantages. In heating a home, for instance, the furnace must supply heat at the same rate that it is lost through the exterior surface. Decrease the exterior surface area, and we decrease the heat loss in proportion, which also decreases the heating requirement. Although it's impractical to build a house in the shape of a sphere, some experimenters have used domes (hemispheres) and cylinders, and even a cube offers a lower heating and cooling requirement than a rectangular shape of unequal sides. The most difficult shapes to heat or cool are those with large numbers of protrusions and additions: bay windows, gables, turrets, and so on. These structural elements increase the surface area significantly without adding a great deal to the total internal volume, and they effectively turn the building into a radiator that transfers internal heat to the surrounding air. Similarly, the best way to keep something cold for the longest period of time is to confine it in a sphere; in fact, this very thing is done in liquefied natural gas tankers and in many other storage tanks designed to hold low-temperature liquefied gases.

Interestingly, this same principle also applies to uncontained fluids. Blow a bubble into a glass of water, and it assumes the shape of a sphere as it rises to the surface; blow a soap bubble and it likewise assumes a spherical shape. Even a raindrop is roughly spherical, distorted only slightly by the effect of aerodynamic drag as it falls through the atmosphere. In all these cases, we have a pair of dissimilar fluids in contact, and forces of surface tension and/or hydrostatic pressure that act to reduce the surface area of contact. In an underwater bubble, the pressure of the water pushes in on the volume of air until it assumes the shape of a sphere and its surface area can be reduced no further. In a soap bubble, intermolecular forces pull the soap film into the minimum area that will enclose the fixed volume of air.

It is not just a fact of abstract geometry, but a principle of nature as well, that the shape of minimum surface area is the sphere.

If we travel into the vacuum of space, where liquid water cannot exist and any soap bubble would explode at the speed of sound, we still find spheres and circles. Here, the governing agent is the force of gravity. Figure 6.3 shows the planet Saturn, which is a huge ball of gas surrounded by rings composed of billions of chunks of ice. Although we generally think of gases as being quite thin, a large amount of gas will generate sufficient gravity to pull it together into a dense sphere of minimum surface area. In the case of Saturn, this sphere is nearly ten times the diameter of planet Earth, and nearly a hundred times as massive.[2] There are several theories for the origin of the rings of Saturn, but all draw upon the fact that ice is not as strong as rock. At the altitude of these rings, a rocky planet would survive very well, but a large ball of ice would be broken into chunks by Saturn's tidal forces. Saturn's rings give us one more example of circles arising from natural forces acting on natural objects.

All of the planets are nearly spherical, and so are all of the larger planetary moons. I say nearly spherical, because inertial (centrifugal) effects do cause rotating bodies like Earth and the sun to bulge at the equator, distorting them into the shape of an oblate spheroid. For Earth, this effect is small: the equatorial diameter is 12,722 km, while the pole-to-pole diam-

Figure 6.3. The planet Saturn. The internal gravity of the planets draws them into approximately spherical shapes.

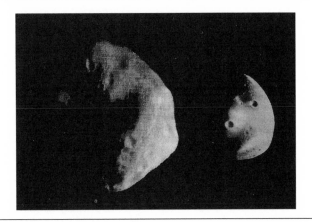

Figure 6.4. Phobos and Deimos, the two moons of the planet Mars. Small moons and asteroids frequently have distinctly nonspherical shapes.

eter is but 34 km shorter. For most purposes, the planets and their larger moons can be considered to be spherical.

Some of the smaller planetary moons, on the other hand, are definitely nonspherical. Figure 6.4 shows Phobos, a potato-shaped hunk of rock measuring about 27 km by 19 km, that orbits Mars once every eight hours, and its smaller sister satellite, Deimos. They are not massive enough to pull themselves into a spherical shapes, yet their rock is too strong to be torn apart and dispersed into a ring by the relatively weak Martian tidal forces. Planetary satellites like Phobos and Deimos, however, are exceptions to the broader pattern we find in the solar system. For the most part, on even the very largest scales, we find that the matter of the universe tends to assemble itself into roughly spherical and circular geometries.

CHAPTER 7

CELESTIAL ORBS

In cultures around the globe, early astronomers and calendar makers took for granted that the sun, moon, planets, and stars orbited a stationary Earth. Such a conventional wisdom was understandable when the world was huge beyond imagination and the points of incandescence in the night sky were puny in comparison. It would make no sense for a massive object like Earth to orbit an apparently small object like the sun or the moon, and most early thinkers probably rejected such a notion without even seriously considering it.

By around 600 B.C.E., a number of Greek thinkers began to defy traditional thinking in their speculations on the connections between humankind and the physical universe, and by 500 B.C.E., the Pythagoreans were teaching not only that Earth is spherical, but that it moves. A generation or two later, Anaxagoras deduced that the sun is much more massive than Earth, and on this basis he argued that our planet must orbit the sun rather than vice versa.

There was nevertheless a great deal of social resistance to any theory that relegated Earth to an inferior role in the grand scheme of the universe. The astronomer Eudoxus, who died in 347 B.C.E., argued that the motions of heavenly bodies could be explained by placing Earth at the center of everything. In Eudoxus's model, Earth was surrounded by a set of concentric transparent spheres (Fig. 7.1), each rotating at its own rate on its own axis. The inner sphere carried the moon, the next carried the planet Venus, and then came additional spheres for the Sun, Mars, Jupiter, and Saturn, all enclosed within a giant outer sphere that carried all of the stars of the

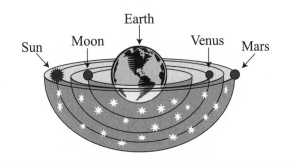

Figure 7.1. Eudoxus's geocentric model of the universe.

constellations. Eudoxus's outer sphere rotated about the poles of the globe, while the others were all inclined to the polar axis by different amounts.

From the beginning, Eudoxus's geocentric model left numerous observations unexplained. For one thing, if the sun's sphere rotates about a fixed Earth once a day, then what makes this sphere wobble north and south 23.5° to bring about the annual solstices, and, more mysteriously, why should this wobble be synchronized with the motion of the outer star sphere in such a way that certain constellations (e.g., Orion the hunter) become visible only in the winter, while others are visible only in the summer? Beyond this fundamental shortcoming, Eudoxus's model also did a poor job of predicting eclipses, failed completely at explaining why the planets occasionally reverse direction against the background of the constellations, and said nothing at all about comets.

These deficiencies might have been enough to doom the geocentric theory to an early demise, were it not for Aristotle (384–322 B.C.), who supported Eudoxus's model. Aristotle's political connections were such (he was tutor to Alexander the Great) that few contemporaries dared to question, let alone attack, his teachings. All of Aristotle's writings were shipped from mainland Greece to Egypt for deposit in the great library at Alexandria, where they were copied and translated into other languages, most notably Arabic and Latin. From Alexandria, these translations were widely disseminated long before the library was destroyed in 415 C.E. As a result of their widespread accessibility, Aristotle's writings became the authoritative source on the workings of the universe for nearly eighteen hundred years.

Aristotle based his science on the premise that the only way to learn the highest truths of the universe is through pure logic. Once an Aristotelean proposition was "proved" logically, it was beyond further dispute, and it would stand for all time. The problem with this point of view is that logic is a uniquely human invention, which the physical universe is under no obligation to obey; in the real world, we are reminded over and again that nature alone has the power to decide what is true or false. Mathematical logic, for instance, tells us that a magnitude 7.0 earthquake centered 100 km offshore cannot possibly trigger a significant tsunami. Yet in 1992, just such an event happened off the Pacific coast of Nicaragua, and the un-anticipated 33-ft sea waves claimed 170 lives and left 13,000 homeless. Logic itself can never give final answers in the real world; at best, analytical reasoning is a pathway to conditional propositions that must be tested further through controlled observation.

Yet in Aristotle's worldview, logical conclusions were expected to stand forever as absolute and immutable truths. That Aristotle himself believed this is of no great importance; the problem was that his writings convinced many others to think this way.

Ptolemy's Epicycles

It's one thing to construct a geometric model of the dynamics of the universe, but a significantly greater challenge to elevate such a model to a theory. A theory must account not only account for past observations, but for future observations as well. Theories must be predictive: to succeed, they must generalize the past and inform us about the future. In both its historical and predictive fidelity, Eudoxus's model was fuzzy at best, and it fell considerably short of qualifying as a theory by any modern standard.

Around 130 B.C.E., Hipparchus of Rhodes pulled together all the astronomical data he could find in existing sources from previous centuries, and he compiled an atlas of about 850 stars. Included in this compilation were data on the motions of the "wanderers": the five unusual stars that refused to stay stuck in any particular constellation, but instead moved in complicated fits of forward and reverse motion within a narrow band that complements the sun's path across the daytime sky. Today, we recognize these wanderers as the five visible planets: Mercury, Venus, Mars, Jupiter, and Saturn. By Hipparchus's time, the observational data were reasonably accurate, having been recorded by hundreds of astronomers pointing giant protractors into the night sky over many centuries. In fact, the measurements were so good that Hipparchus discovered that the north celestial pole (the axis of Eudoxus's outer sphere) wobbles about a bit, and that from time to time over many centuries, different stars become the North Star. This is surely one of the most amazing discoveries of ancient times, for we now know that the period of the precession of Earth's axis is 26,000 years! Hipparchus discovered the precession of the equinoxes using data that spanned barely one percent of this period.

Around 150 C.E., Claudius Ptolemy took on the problem of developing an improved model of the universe. Working in Alexandria, Ptolemy was disturbed by the discrepancies among the observational data, Aristotle's fundamental "truth" that the universe is Earth-centered, and the inaccuracies and omissions of Eudoxus's model. He realized that he couldn't blame the data, which had been independently corroborated by many observers, and he chose not to dispute Aristotle. To Ptolemy, it was Eudoxus's model that had to be corrected. Planets couldn't just be stuck to transparent rotating spheres.

What other geocentric models might work? One possibility was a wheel that rolled along a sphere. It was well known to the mathematicians of Ptolemy's time that a point on the rim of a rolling wheel travels forward at a varying speed, and that in doing so it traces a curve called a *cycloid* (Fig. 7.2). A variety of related motions are possible if the point is not on the wheel's circumference, but instead either inside the wheel or on an extension of its radius; such curves are usually called *trochoids*. If, on the

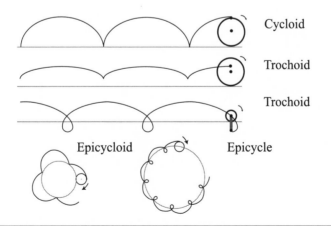

Figure 7.2. Curves traced by a fixed point attached to a rolling wheel.

other hand, a wheel rolls around a circle, a point on its circumference traces a curve called an *epicycloid*. This suggested that an epitrochoid (a trochoid on a circle) might be the kind of curve Ptolemy needed to explain the retrograde motion of the planets. The numbers, however, just couldn't be made to work without allowing the rolling circle to slip as it rolled. Ptolemy finally defined a new curve, the *epicycle*, which is traced out by a point on the circumference of a rotating circle as its center follows the circumference of a larger circle. In an epicycle, the circle's rotational rate and its forward speed are independent, and can be specified independently.

By adding properly sized epicycles to Eudoxus's spheres and adjusting their speeds and rates of rotation, Ptolemy was able to devise a geometrical model that came reasonably close to predicting the actual planetary motions. Where discrepancies still existed, Ptolemy added smaller rotating circles that traced out epicycles upon some of the prior epicycles. Eventually, by adjusting some forty-odd numerical parameters, he succeeded in getting his geocentric model of the universe to conform to all of the documented astronomical observations of his time. A simplified diagram of Ptolemy's model is shown in Figure 7.3.

Ptolemy included his geocentric theory with its epicycles upon epicycles in a monumental thirteen-volume work titled *The Mathematical Collection*, which eventually reached Europe via the Arabs as the *Almagest*. It's important to note that Ptolemy's revised geocentric theory was far from sloppy mathematically; in fact, most people who take the time to understand it find it quite impressive. And it actually did a fairly good job of predicting

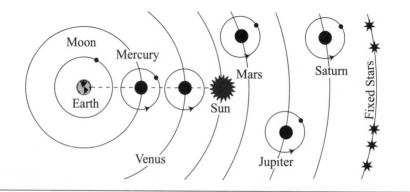

Figure 7.3. Claudius Ptolemy's geocentric model of the universe.

the positions of the planets, the stars, the sun, and the moon on any particular day of any year, at any date into Ptolemy's foreseeable future.

Science and Authority

After the fall of the Roman empire, most of Europe fell into the Dark Ages, while learning continued to prosper in much of the Arab world. In western Europe, the Roman Catholic church became the principal guardian of intellectual knowledge, which also put it in the position to translate, interpret, and censor any writings, current or past. Sometime after the ninth century, courtesy of the Arabs, Ptolemy's *Almagest* found its way into a number of European monasteries, as did many of the writings of Aristotle. In 1273, the Dominican friar Thomas Aquinas completed his *Summa Theologica*, another monumental multivolume work that "proved" most of the teachings of the church in the language of Aristotelian logic. Aristotle's philosophy had a particular appeal to Aquinas because once one proved a conclusion through this system, there remained no mechanism for appeal. Aristotle's logic was self-contained, and truth was truth, immutable and unquestionable. The papacy agreed, and endorsed the writings of Aquinas, with their Aristotelian philosophical basis, as official church dogma.

But in accepting the restructuring of its teachings in the language of Aristotelian logic, the Catholic church also took on some baggage. Aristotle's conclusion that Earth is the center of the universe now became a religious truth, and for anyone to claim otherwise was a heresy. Even before Aquinas, in 1233, Pope Gregory IX had formally established the Inquisition, which was a special tribunal to stamp out all heresy. After Aquinas, thinkers were punished with equal severity for claiming that Jesus was not divine or for claiming that Earth is not the center of the universe. And for those curious enough to wonder just *how* an Earth-centered universe could possibly be consistent with what one sees happening in the night sky, there was Ptolemy's *Almagest* to explain the mathematical details of the epicycles upon epicycles.

Science is based on ignorance; it thrives only by exploring unanswered questions, and letting the chips fall where they may. When a society or religion proclaims it already has all the answers, there can be no science. And so, until the mid-1500s, the Aristotelian worldview and Ptolemy's geocentric theory stood virtually unchallenged in church-dominated Europe.

The tragedy of this period in history is that there were undoubtedly many thousands of fertile minds that could have accomplished great things had they not been conditioned to accept that there were certain ideas and concepts they shouldn't question.[1]

Toward a Heliocentric Theory

In science, as well as in other investigative disciplines (e.g., medicine and law), it often happens that a given set of observations will support several competing explanations. This raises the question of how we might choose a "best" theory from several alternatives. As early as 1340, the English cleric and philosopher William of Occam addressed this question by proposing a principle of parsimony that has come to be known as "Occam's razor": When several conflicting explanations are proposed for the same set of observations, the best explanation is the one that depends on the fewest independent assumptions.[2] Occam did not expect inductive logic to reveal truths in any absolute sense, yet he couldn't accept that all explanations might be of equal merit. His principle of parsimony therefore does not refer to "truth," but rather to the "best" explanation—the "best" being that which explains the most in terms of the least.

On November 1, 1755, for instance, the city of Lisbon, Portugal, was struck by an earthquake, then a tsunami, and finally a great fire that burned for several days and consumed everything that hadn't previously been destroyed. Around forty thousand people died. One explanation for this disaster might be that Lisbon just had a particularly bad day, with a fire and a tsunami and an earthquake all striking within a few hours of one another. A second explanation is that all of the devastation resulted from a single source: an earthquake that took place offshore, under the sea. In the second scenario, earthquake waves raced through the seabed and into the city, collapsing many buildings into piles of splinters and overturning stoves (which would have been lit at that time of the year). The same earthquake also jostled the sea, sending great waves into the city's waterfront an hour or so later (because water waves travel slower than earthquake waves). By the time the tsunamis receded, the numerous small fires in the higher sections of the city had time to combine into a roaring conflagration that, fueled by the splinters, could not be controlled. In this scenario, the forty thousand victims died not from three separate events, but from a single event that triggered three devastating effects. But which explanation is true? Nothing in Aristotelian logic tells us how to decide. Yet the second explanation, with its fewer independent assumptions, is surely more probable and more compelling.

Ptolemy's geocentric theory, with its forty independently adjustable numerical parameters, was not an impossible explanation for the observed motions of celestial bodies. It was, however, somewhat implausible. Ptolemy's model was rendered even less compelling after Mikolaj Kopernik, a sixteenth-century Polish cleric and astronomer who wrote under the latinized name Copernicus, developed an alternative theory that required only fifteen independent numerical quantities to explain the historically accepted astronomical observations.

Copernicus accomplished this feat by transfering the center of the universe from Earth to the sun. He then treated Earth as just another of the planets, along with Mercury, Venus, Mars, Jupiter, and Saturn. In this heliocentric, or sun-centered, scheme, the planets all orbit the sun in circular paths (Fig. 7.4), the inner planets moving faster than the outer ones. No need for Ptolemy's epicycles upon epicycles; all it took was a simple set of concentric circles, one for each planetary orbit. This austere model satisfactorily explained retrograde planetary motion: as planet Earth overtakes the slower outer planets, Earth-bound observers are treated to the illusion that these planets reverse their motions relative to the more distant star field. Copernicus was careful to point out that his heliocentric theory should not be viewed as "truth," but rather as a simplified way to compute the geometrical relationships among moving planets. Even so,

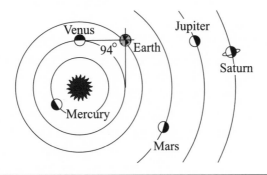

Figure 7.4. Copernicus's heliocentric model.

given that any implication of a heliocentric universe was sure to be controversial in his church-dominated culture, Copernicus delayed publication of his book for some thirteen years, until he lay on his deathbed in 1543.[3]

Kepler's Laws

Johannes Kepler (1571–1630) was driven by the belief that the structure of the universe is geometrical, not just approximately, but as exactly as human senses are capable of observing. Kepler was an excellent computational mathematician and deeply religious, and his life's ambition was to discover that perfect theory of the universe that would allow him to peer into the mind of its Creator.

Although influenced by the writings of Copernicus, Kepler was initially skeptical about the validity of the Polish cleric's heliocentric model. In referring to the available observational data, he noted that Ptolemy's geocentric model actually did a better job of predicting the positions of the planets. In particular, the Copernican model's nest of concentric circular orbits failed to explain why the planet Mercury sometimes reverses its motion when it is 18° from the sun (as viewed from Earth), but other times this reversal is delayed until Mercury is some 27° from the sun. Circles are symmetrical, and if they are also concentric, motions along such circles should be quite regular. The inaccuracies of predictions from Ptolemy's model were modest by comparison.

Yet Kepler found the Ptolemaic model, with its dozens of independently adjustable parameters, an unaesthetic starting point for building an improved geometrical model of the universe. Yes, he could arbitrarily add more epicycles upon more circles, introducing new variables that could be adjusted to conform ever more precisely to the astronomical observations, yet as a religious man he couldn't believe that the Creator would have actually designed the universe in such a clumsy manner. The true structure of the universe, Kepler felt, must be elegantly simple.

The number of direct astronomical observations one might hope to make in a human lifetime is fairly limited. If today Earth is at its point of closest approach to the planet Mars, some 778 days must pass before the next time this happens. For the planet Jupiter, the corresponding waiting time is about 12 years, and for Saturn it is close to 30 years. Although the planet Neptune was discovered in 1846, it has yet to complete one full orbit around the sun; it is projected to do so in the year 2011.

Kepler desperately needed a comprehensive set of reliable and accurate measurements of planetary positions on various dates, data he had neither the time nor the visual acuity to collect himself. In the year 1600, he traveled to Prague to the observatory of the wealthy and eccentric astronomer Tycho Brahe (1546–1601), who for nearly three decades had kept meticulously precise data of planetary positions. Although the telescope was yet to be invented, Tycho's observations consistently approached the two-arc-minute limit of angular resolution of the naked human eye. Tycho had always hoped to use his reams of measurements to structure a new theory of the universe, but he found that his own mathematical skills were not up to the task, and in any case he became increasingly distracted by extravagent parties. Although he realized he needed Kepler's mathematical skills to build his theory, he was nevertheless reluctant to simply hand over his life's work to a young upstart. As a result, Tycho and Kepler were not particularly compatible partners, Tycho doling out his data in only small scraps and Kepler becoming increasing annoyed at the bedlam created by the constant string of guests at the observatory.

When Tycho died the following year, Kepler gained access to the full collection of thirty years' worth of observational data. Initially, Kepler continued to work on Tycho's theory, a geocentric-heliocentric hybrid in which Earth was immovable, the sun revolved around Earth, and the five planets revolved around the sun. Ultimately, however, the best Kepler could accomplish with Tycho's model still resulted in a discrepancy of eight arc-minutes in some of the Martian data. Despite their prior conflicts, Kepler retained such respect for the integrity of Tycho's data that he rejected the Tychonic model on the basis of this tiny discrepancy. "Upon this eight minutes," he wrote, "I will build my universe."

At this point, Kepler went back to the heliocentric model of Copernicus and sought to improve it. Although the Copernican model appears simple enough in its standard representations, mathematical complications arise because we humans do not see the solar system from a fixed point in outer space, but rather from the surface of a moving Earth. In an age before calculators and computers, the challenges of calculating the position of a moving object as observed from a moving platform were considerable, particularly given that planetary objects and our Earth-based platform

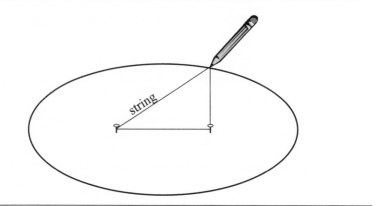

Figure 7.5. Drawing an ellipse.

might move at variable speeds along paths that are not circles. Kepler, however, rose to this challenge, and by 1609 he formulated the following principle:

> *Kepler's first law:* The path of each planet is an ellipse, with the sun at one focus.[4]

An ellipse is defined as the locus of points for which the sum of the distances from two fixed points (the foci) has a constant value. Figure 7.5 shows how this definition can be used to draw an ellipse: using two thumbtacks to define the foci, we tie a piece of string into a loop and drop it over the thumbtacks, then pull the loop tight with a pencil point and draw the closed curve. Clearly, a circle is an ellipse whose two foci coincide. As the distance between foci increases, an ellipse becomes more eccentric, and its deviation from a circle grows more apparent.

It takes two parameters to describe the size and shape of an ellipse. Kepler chose to use the ellipse's semimajor axis and its eccentricity (Fig. 7.6). The *semimajor axis* was a natural choice for one parameter, because as an ellipse approaches the shape of a circle, the length of its semimajor axis approaches the radius of that circle. Kepler's second parameter, the *eccentricity*, specifies how far the foci are displaced from the center of the ellipse; numerically, the eccentricity is the ratio of this offset distance to the length of the semimajor axis. A circle has an eccentricity of zero (i.e., no offset), whereas a long narrow ellipse has an eccentricity that approaches one.

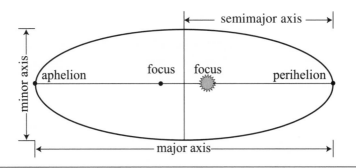

Figure 7.6. The ellipse as applied to a planetary orbit. The sun is always at one focus. The eccentricity of the orbit is the distance between the ellipse's center and a focus, divided by the length of the semimajor axis.

Using the wealth of Brahe's data now at his disposal, Kepler found that all the observational data on the motion of Mars were consistent with an elliptical orbit whose semimajor axis was 1.52 times that of Earth, and whose eccentricity was 0.093. He worked out similar orbits for the other known planets, finding, for instance, that the eccentricity of Mercury's orbit was quite large: 0.206. Modern values, including figures for the more recently discovered planets, are given in Table 7.1. Kepler was quite encouraged, for his elliptical orbits successfully explained many specific observations, including the fact that Mercury sometimes reversed when only 18° from the sun, while other times reversing at 27°. These reversals were a direct consequence of the relative positions of Earth and Mercury on their respective elliptical orbits when they and the sun happened to form a right triangle.

Kepler's first law established *where* the planets might stand in relation to one another and to the sun, but not *when* they might be found in these positions. Ptolemy had assumed that all the planets moved along their epicycles at constant angular speeds, with the epicycles themselves following circular orbits at (other) constant speeds. Kepler found that Ptolemy's assumption of constant speeds was inconsistent with his newly discovered elliptical orbits. Instead, the planets seemed to speed up and slow down in different parts of their orbits. Kepler began to look for another geometrical pattern that might describe this effect.

As a planet travels along an ellipse, its distance from the sun obviously varies. The point of closest approach to the sun, which lies along the major axis, is called the planet's *perihelion*. The point of farthest distance from

Table 7.1. Data on the sun and planets.

Object	Diameter (km)	Rotation period (days)	Orbital period (years)	Orbital semimajor axis (AU)	Orbital eccentricity
Sun	1,391,400	25.4			
Mercury	4,878	58.6	0.244	0.387	0.206
Venus	12,104	243.0	0.616	0.723	0.007
Earth	12,756	1.00	1.000	1.000	0.017
Mars	6,787	1.02	1.881	1.52	0.093
Ceres	1,020	0.38	4.6	2.77	0.08
Jupiter	142,800	0.41	11.9	5.20	0.048
Saturn	120,660	0.43	29.5	9.54	0.056
Uranus	50,800	0.72	84.0	19.18	0.05
Neptune	48,600	0.67	164.8	30.07	0.01
Pluto	2,300	6.4	247.7	39.44	0.25

Notes:

1 AU = 1 astronomical unit = length of semimajor axis of Earth's orbit = 149,600,000 km = 92,980,000 miles.

Ceres is the largest of several thousand asteroids between Mars and Jupiter.

the sun is called the *aphelion*. It didn't take Kepler long to notice that the planetary speed at the perihelion is greater than at the aphelion. How much greater? Simply, it turns out, in proportion to the ratio of the solar distances at aphelion and perihelion.

But the perihelion and aphelion are only two points on a planet's orbit; how does the speed vary along the rest of an orbit? This question was more of a challenge, but Kepler already had a starting point to guide him. He noted that if he multiplied a planet's speed (i.e., distance/time) by the radial distance to the sun, the result had dimensions of distance²/time, or area/time. The data had already shown that this quantity is the same at the aphelion and the perihelion, provided that the reference time interval is the same. Was this also the case at other points on the orbit? Kepler checked it out numerically, and it was so. This led to his formulation of his second law, also in 1609:

Kepler's second law: The line between the sun and a planet sweeps through equal areas in equal time intervals.

Although this law is slightly more abstract than Kepler's first law, it was still cast in the language of pictorial geometry. Figure 7.7 shows an ellipti-

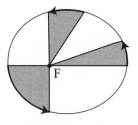

Figure 7.7. Kepler's second law of planetary orbits. In equal times (say, one month), the three areas are equal.

cal orbit with three arcs, each corresponding to the distance some planet moves in an arbitrary interval of time (say, one month). If we know one of these arc lengths, we can calculate the area of the corresponding sector of the ellipse. Kepler's discovery was that such an area is equal to the area of any other one-month sector of the same orbit. Since the shape of the ellipse has already been established, this permits a computation of the planet's speed at any and every point in its orbit, provided only that we know its speed at one reference point of the orbit (say, at the perihelion).

Now computing sectors of an ellipse is not an easy task, even when one is using calculus and some number-crunching techniques that were not known in Kepler's day. To establish his second law, Kepler laboriously developed page after page of tables of areas of elliptical sectors, computed to an approximation at least as good as the accuracy of the astronomical data he was working with. And when he did this, he found that he could indeed predict the positions of planets on any given day to a precision consistent with the resolving limit of the human eye.

As of this date, 1609, Kepler had a successful heliocentric model that required the specification of just sixteen independent parameters: the eccentricities of the six planetary orbits, the semimajor axes of the five extraterrestrial planets relative to the semimajor axis of Earth, and the five velocities at a known reference point in the orbits, relative to the corresponding velocity of Earth. This was a vast improvement on Ptolemy's geocentric model, which required a specification of more than forty parameters, yet which still failed to make predictions with a precision approaching the acuity of the human eye. Further, in Ptolemy's model all these forty-odd parameters were established not by direct computation but by trial and error, that is, by adjusting the size and speed of the epicycles until the result conformed to what was observed from Earth. Kepler, by contrast, established the values for his smaller number of parameters by direct calculation from physical measurements of position and time.

Yet Kepler wasn't satisfied; he wanted to reduce the number of parameters in his model even further. Was there any way to eliminate the need to specify separately a planet's speed and the size of its orbit? Certainly the inner planets move faster than the outer planets, but do they do so according to any predictable pattern?

This problem challenged Kepler for the next ten years; only in 1619 did he publish the answer in his third law, also sometimes referred to as the *harmonic law*:

> *Kepler's third law*: The ratio of the cube of the semimajor axis to the square of the period of revolution is the same for each planet.

One reason it took Kepler so long to arrive at this statement was that it cannot easily be pictured geometrically.[5] Further, while the second law requires a specification of speed, the third law emerges only from considering a planet's orbital period (the time for one complete orbit). Still, knowing the orbital period and the size of the ellipse permits a computation of the average orbital speed, and from this the reference velocities required by the second law can also be established by computation. This eliminated another five parameters from Kepler's theory, so now only eleven numbers needed to be specified independently: six orbital eccentricities and five semimajor axes. In other words, the ellipses themselves now said it all: they defined the path a planet follows, the time it takes to complete an orbit, and the planet's variations in speed at different points on its orbit.

Kepler's theory was by any measure a monumental intellectual accomplishment, and one that was consistent with Occam's razor. The more than forty independent assumptions of Ptolemy's geocentric model had been successfully reduced to but eleven measured parameters in Kepler's heliocentric universe.

Still, Kepler didn't feel he was finished. He wanted to further reduce the number of independent parameters, to perhaps six, and ideally to one. If the workings of the universe ultimately depended on only a single number, this would confirm Kepler's belief that mathematics, and geometry in particular, was the language of the Creator. The single number needed to describe the universe would effectively be the spoken word of God.

Kepler had an idea about how to proceed. One question that had haunted him since his earlier years as a professor at Graz was this: Why six planets? Why not three, or fourteen, or thirty-five? As a geometrician, Kepler had also long been intrigued by the so-called Platonic solids, regular polyhedrons whose faces were identical regular polygons. Since the time of Euclid, it had been well established that there are only five such solids: the tetrahedron (four faces, each an equilateral triangle), the hexahedron (six faces, each a square), the octahedron (eight faces, each an equilateral triangle), the dodecahedron (twelve faces, each a regular pentagon), and the icosahedron (twenty faces, each an equilateral triangle). Every Platonic solid can be enclosed in a sphere that simultaneously touches all its apexes, and every Platonic solid can envelop a smaller sphere that simultaneously touches the center of each polygonal face.

Long before he met Tycho Brahe, Kepler had noticed that if he nested six spheres inside each other, sizing and spacing them according to the five Platonic solids, the result bore a remarkable similarity to the Copernican heliocentric model. When he did the calculations in more detail, however, Kepler found discrepancies between the proportions of these nesting solids and the apparent spacing of the planets. After 1619, his three laws complete, it became apparent to Kepler why the nesting spheres hadn't worked; if anything, they needed to be ellipsoids or some other shape rather than spheres. For the last eleven years of his life, Kepler continued his efforts to use the Platonic solids to establish the spacing of the planets and thereby reduce the number of parameters of his theory to six. He never was successful in this portion of his endeavor. Today we know of nine major planets, thousands of asteroids, and dozens of periodic comets that orbit the sun, some in overlapping paths. A mere five Platonic solids cannot possibly explain their spacings.

Galileo and His Telescope

Kepler conducted most of his work in Protestant regions of Europe, where he was under no pressure to please the pope. Galileo Galilei (1564–1642), on the other hand, lived and worked in Italy, had acquaintances and colleagues in the church hierarchy, and ultimately got into serious trouble by contradicting the cosmological teachings of the church.

Around 1590, Galileo conducted a series of experiments at the University of Pisa, on falling bodies and on balls rolling down ramps. Aristotle had written that such objects attain speeds in proportion to their masses, but Gaileo discovered that this ancient "truth" stood at odds with his own observations. Variations in mass, it turns out, do not affect the speed of a falling body, nor for that matter the speed of a pendulum, all other factors being equal. This finding impressed Galileo, not just because of the specific results, but because it opened the door to the possibility that Aristotle may have been wrong about much more. Galileo went on to discover that the trajectory of a projectile near Earth's surface is a parabola, and he corresponded with Kepler regarding ideas about Copernicus's heliocentric universe.

Then, around 1609, Galileo received reports from the Netherlands of a an optical instrument that magnified distant objects. Galileo built one for himself, improved on it, and aimed it into the night sky. The moon, to his amazement, was not a perfect sphere, but had mountains and valleys. The planet Venus was not a point of light, but rather a disk that went through a series of phases just as the moon did. The planet Saturn was oval in shape (Galileo's telescope could not resolve the rings). Using the telescope to project the sun's image onto a screen, Galileo found the sun to be pockmarked with spots that moved from day to day. Most surprising, however, was the planet Jupiter: not only was it a disk rather than a point of light, but it had four satellites orbiting it. The periods of these Jovian satellites ranged from 1.8 days to around 17 days, the inner satellite moving the fastest and the outermost the slowest, very much like a tiny Copernican solar system. In 1611, Galileo visited the papal court in Rome and demonstrated his telescope, allowing priests, cardinals, and visitors to confirm what he had discovered.

To Galileo, these observations were indisputable evidence that Ptolemy's geocentric model was wrong. The observed phases of Venus could be accounted for only if Venus orbits the sun rather than Earth, and the Jovian moons (unknown to Ptolemy) obviously did not orbit Earth. To Galileo, the evidence was indisputable that Earth does not stand immovable at the center of the universe. The Catholic church, however, chose to differ, claiming that observations of the planets are not in themselves a theory of the planets. To the church, faith had a higher value than direct observation, and if there were logical contradictions between the two, it was the observational evidence that had to be ignored. The church insisted, Galileo's new observations notwithstanding, that Earth was still the center of everything.

The heliocentric theory, however, continued to attract believers, particularly after Kepler published his first two laws in 1609. In 1616, the pope officially denounced the Copernican system (and, by implication, all heliocentric theories) as injurious to faith. Galileo was summoned to Rome and was warned not to teach or uphold the Copernican theory under pain of excommunication.

Galileo in fact heeded this warning for several years, but the issue kept gnawing at him. In 1632, he published his *Dialogue on the Two Chief World Systems*, in which Simplicius, a thinly veiled caricature of Pope Urban VIII, stupidly and ineffectively attempts to defend the geocentric theory.[6] For this, Galileo was again summoned to Rome. This time he was arrested, tried by the Inquisition, and under threat of torture and death was forced to recant all his teachings and writings that suggested a moving Earth. The court spared his life, and he lived his remaining nine years under house arrest. A portrait painted two years before his death in 1642 carries the inscription *E pur si muove* (nevertheless, it does move), suggesting that to the end Galileo remained intellectually defiant. He was pardoned by Pope John Paul II in 1992.

CHAPTER 8

FROM CONICS
TO GRAVITY

Slice a right circular cone parallel to its base, and the section is a circle; run a slice a bit obliquely to the base, and the section is an ellipse. When such slices are made at increasing angles, as shown in Figure 8.1, the eccentricities of these ellipses increase. At an eccentricity of 1, the ellipse ceases to close on itself, and the section is a parabola. With further increases in the angle of the slice, the resulting curve is a hyperbola. Although at first glance a hyperbola may appear very similar in shape to a parabola, the difference is that the outer extensions of a hyperbola become indistingishable from two diverging straight lines, called *asymptotes*. A parabola does not have such asymptotes.

The Greek mathematician Apollonius of Perga (247–205 B.C.E.) described these conic sections in the formal language of Euclid's geometry and coined the terms *ellipse*, *parabola*, and *hyperbola*. He also established that this general class of curves includes as special cases the geometrical point (at the apex of the cone) as well as pairs of straight lines (degenerate hyperbolas passing through the cone's apex). Apollonius did not, however, see much connection between conic sections and the physical world. Although straight lines and circular geometries do exist in nature (at least to reasonable approximations), ellipses seemed to arise only as illusions brought on by viewing a circle obliquely, and parabolas and hyperbolas seemed to have no natural counterparts at all.

Prior to the early 1600s, the trajectories of cannonballs were approximated by rising and falling straight-line segments, or else by a pair of such segments connected by a semicircular arc. With Galileo's discovery that

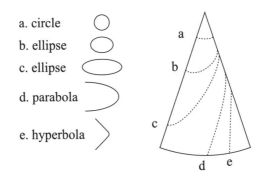

a. circle

b. ellipse

c. ellipse

d. parabola

e. hyperbola

Figure 8.1. The conic sections. Circles, ellipses, parabolas, and hyperbolas can be generated by running a plane through a cone.

projectiles follow parabolic trajectories, it became apparent that the parabola is a very common curve in the physical world; in fact, we have all seen streams of water from fountains tracing this shape (Fig. 8.2). Shortly thereafter, Kepler showed that the ellipse also arises in nature, in the orbits of planets around the sun. Thus, by 1609, three of the four types of conic sections had been associated with "natural" curves, and only the hyperbola remained curiously ignored by the laws of nature. This situation would be resolved some sixty years later, courtesy of Isaac Newton. It turns

Figure 8.2. Streams of water in fountains trace parabolic curves in space.

out that all four sections—circle, ellipse, parabola, and hyperbola—are indeed ubiquitous in the natural world.

The Rise of Analytic Geometry

Although the invention of the familiar x-y coordinate system is often attributed to the mathematician and philosopher René Descartes (1596–1650), systems of rectangular coordinates were in fact used by the ancient Egyptians thousands of years earlier to reestablish land boundaries after the annual spring floods, and to preserve geometric proportions when rendering large-scale public art. Ancient maps drawn by Eratosthenes used a system of longitude and latitude that, even over sizable distances, was effectively a Cartesian coordinate system. It has been speculated that humans arrived at the idea of using a reference frame of two mutually perpendicular axes by observing nature: the surface of the sea is horizontal, whereas the string of a plumb bob is vertical, and these two "natural" orientations are perpendicular.[1] It takes no major conceptual leap to lay such a system flat, superimpose two linear scales, and use it as a reference frame for mapping or surveying.

What Descartes *did* accomplish was to use rectangular coordinates in a new way, bridging the gaps between formal geometry and algebra.[2] Algebra, up to his day, was concerned only with finding unique numerical solutions to sets of equations. To solve for one unknown required one equation, to solve for two unknowns required a set of two independent equations, three unknowns called for three equations, and so on. Solutions to algebraic equations, when they existed, were fixed and static; equations, in other words, were not used to describe how one solution might evolve into a new solution, as with the position of a moving planet. Descartes, however (as well as his contemporary, Pierre de Fermat), recognized that algebraic equations could be meaningful even when they did not yield unique solutions. A single equation in two unknowns, for instance, although it cannot be solved uniquely, may be viewed as a set of *possible* solutions that define a curve in a plane. Similarly, an equation in three unknowns defines a surface in space. In analytic geometry, algebraic unknowns are no longer treated as fixed quantities; instead, they are *variables* that may take on a whole range of values.

The equation $(x - 2)^2 + (y - 5)^2 = 36$, as an example, yields a set of coordinate pairs (x, y) which, when plotted on a Cartesian plane, traces out a circle whose center is at the point $x = 2$, $y = 5$, and whose radius is six units. The equation $5x^2 + 7y^2 = z$, on the other hand, yields a set of co-ordinate triplets (x, y, z) that define a surface in three-dimensional Cartesian space; in this case, the surface is an elliptic paraboloid. (Viewed along the z-axis it has an elliptical cross section, while viewed along the other axes it has a parabolic cross section.) By the 1640s, many European mathematicians were exploring the wonders of this new *analytic geometry*, and discovering a broad variety of geometrical curves and surfaces that can be described by algebraic equations.

Descartes himself was especially interested in conic sections, and he quickly discovered that all curves in this class can be described by algebraic equations of very similar form (these equations are listed in Appendix C). Further, given the equation of any specific conic section, it was a fairly easy matter to compute every possible geometrical parameter of the curve (e.g., the center or apex, the lengths of the major and minor axes, the coordinates of the focus or foci, the eccentricity, and the slopes of the asymptotes). For the first time in history, one could now *read* an equation, and if it fell into one of the recognized forms, one could immediately sketch the curve with a minimum of computation. Conversely, one could begin with a particular curve, and if it had a recognizable geometry, translate it immediately into an algebraic equation. With the advent of analytic geometry, geometry and algebra were merged into a single coherent language.

Other Coordinate Systems

There is no reason why curves *must* be described in terms of rectangular coordinates (x, y). In the physical world, a parabolic stream of water pays no attention to whether or not the observer imposes a Cartesian reference frame on it. Similarly, in the mathematical world, a geometric curve must retain its shape whether we describe it in terms of Cartesian coordinates or through some alternative system for locating points on a plane.

What other coordinate systems are possible? An infinity of them. For practical reasons, however, we usually insist on a requirement of *orthogonality*, which means that for each point on a plane, the two coordinate directions always meet at a right angle.[3] Still, there are many ways to meet this requirement. Next to the rectangular system, the most common is the polar coordinate system. Here, each point on a plane is specified by the coordinate pair (r, θ), where r is the radial distance from an origin, and θ

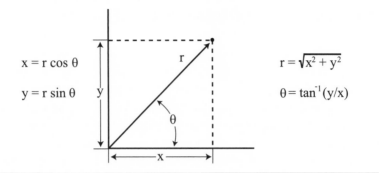

Figure 8.3. Rectangular and polar coordinates.

is the angle measured from a 0° reference line. Figure 8.3 shows how a given point may be located by using either rectangular or polar coordinates. Today, even the most basic scientific calculators perform numerical rectangular-to-polar conversions with a built-in function key.

Descartes and his contemporaries recognized that the algebraic description of a geometrical curve can sometimes be simplified by using polar rather than rectangular coordinates. As an example, there is a curve called a *cardiod* (Fig. 8.4), so named because it resembles a heart. Geometrically, this curve is generated by rolling one circle around the circumference of a second circle of equal radius. If we designate the radii of these circles as a, the area bounded by the cardiod is $3\pi a^2/2$, and the total arc length is simply

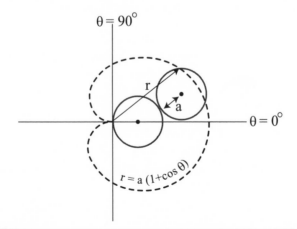

Figure 8.4. The cardiod is one of many shapes more easily described in polar than in rectangular components.

8a. Yet in Cartesian coordinates, points (x, y) on the cardiod must satisfy the equation

$$a^2y^2 = x^4 - 2ax^3 + 3x^2y^2 - 2axy^2 + y^4,$$

which is a mouthful for anyone to read, not to mention being a clumsy equation for computing values of y from values of x. Alternatively, however, the same cardiod can be described in polar coordinates (r, θ) through the equation

$$r = a(1 + \cos \theta).$$

This not only looks simpler, but it *is* simpler for computational purposes. This second equation also preserves a direct link to the way the cardiod was generated from the one circle rolling on the other, for if $\theta = 0$, then $r = 2a$ (the obvious x-intercept of the curve), and if $\theta = \pi$ radians, then $r = 0$ (the x-value of the cusp).

Other orthogonal two-dimensional coordinate systems are sometimes useful. One example is the elliptic coordinate system (Fig. 8.5), which may be viewed as intersecting families of confocal ellipses and hyperbolas. Notice that these ellipses and hyperbolas always intersect at right angles. In this system, any point P can be located by the pair of coordinates (u, v), where u identifies the particular ellipse running through the point and v identifies the hyperbola running through that point.

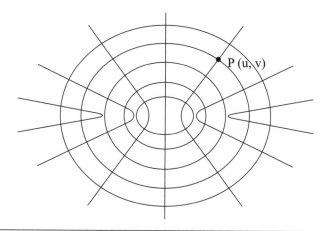

Figure 8.5. Elliptic coordinates. This is but one of many nonrectangular yet orthogonal coordinate systems.

There is also a variety of three-dimensional orthogonal but nonrectangular coordinate systems. The most common are cylindrical polar coordinates (r, θ, z) and spherical polar coordinates (r, θ, ϕ), but other systems such as paraboloidal coordinates, oblate spheroidal coordinates, and toroidal coordinates can also be useful at times. Switching the coordinate system never alters the curve or the surface geometry itself; the coordinate system does, however, affect the form of the equation used to describe any given curve. Many problems that are intractable when described with rectangular coordinates become manageable through a creative change of coordinate system. We will see shortly that this idea was of great value to Isaac Newton in helping him analyze how gravity might be responsible for the observed motions of the planets.

Conics and Optics

It was in the field of optics that the new principles of analytic geometry first found practical application. Since ancient times, jewelers and weavers often magnified their work by examining it through a glass globe full of water.[4] Even if the glass was pure and the globe was sufficiently spherical, however, only the center portion of the image was sufficiently distortion free to be of use (Fig. 8.6). Toward the outer portions of the image, straight lines became so curved that the geometry of the magnified object was hopelessly muddled. Today, this effect is referred to as *spherical aberration*.

In the thirteenth century, it occurred to Roger Bacon that if only the center of the image was distortion free, then most of the globe could be

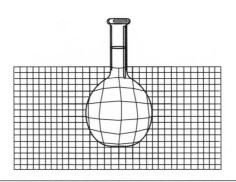

Figure 8.6. Using a spherical flask of water as a magnifier. The distortion is called *spherical aberration*.

eliminated. In experimenting with this idea, he discovered that even the water could be eliminated; to create a lens, all he needed was a solid piece of glass whose opposite surfaces were sections of spheres. This insight led to the invention of the first crude eyeglasses, fabricated by pouring molten glass into curved molds. The actual shape was difficult to control because of thermal expansion and contraction, and the lenses so made often had bubbles and other optical imperfections. In the seventeenth century, however, it became standard practice first to manufacture the glass and concentrate on its quality, then, after it cooled and solidified, to pick the clearest pieces and grind their surfaces into the proper curvature. This led to a vast improvement in the quality of eyeglasses and other lenses, and permitted the invention of the telescope and microscope. Of course, the need for grinding also increased the cost of these improved lenses.

Still, the problem of geometrically distorted images persisted, and this was particularly troublesome in telescopes of increased magnifying power. The theorems of analytic geometry suggested that this problem might be a result of the inherent geometry of the sphere, which was used as a template for grinding the curved lens surfaces. Although the hyperboloid (a surface created by rotating a hyperbola about its axis of symmetry) was shown to be a more appropriate shape for the surfaces of a lens, a hyperboloid turns out to be a difficult shape to grind accurately.[5] One can check the geometrical integrity of a spherical surface simply by noting that the radius of curvature must be the same at every point (easily verified with a simple instrument called a spherometer); for a hyperboloid, however, the radius of curvature is smaller for points near the axis of symmetry, and larger for points nearer the circumference of the lens. There is no easy way to grind a piece of glass into a hyperboloid, nor to confirm through measurement that an acceptable precision has been achieved.

Lenses also present other difficulties that limit the quality of their images. Even if spherical aberration is completely eliminated, an image of a planet viewed through a lens will nevertheless be surrounded by a multicolored halo—a defect known as *chromatic aberration*, which results from the fact that glass refracts different colors at slightly different angles. Although chromatic aberration can be reduced by introducing additional compensating lenses into an optical system, such an approach is practical only with small eyepieces. The bigger the lens, the more prominent its defects, and the more difficult it becomes to correct aberrations by stacking a series of lenses together.

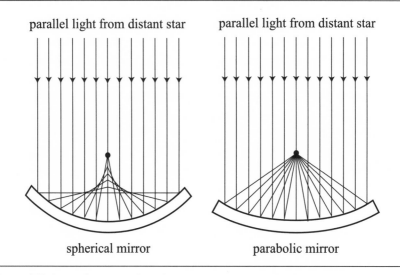

parallel light from distant star   parallel light from distant star

spherical mirror   parabolic mirror

Figure 8.7. Image formation by spherical and parabolic concave mirrors.

The image quality of a telescope can be greatly improved by avoiding large objective lenses and instead focusing the light with one or more curved mirrors. Because light does not travel *through* a mirror, there can be no chromatic aberration. And because light travels in straight lines, image formation on reflection can be predicted through analytic geometry alone. It is easy to demonstrate, for instance, that a concave spherical mirror focuses distant light to multiple points and that this results in spherical aberration similar to that observed with spherical lenses (Fig. 8.7). A concave parabolic mirror, on the other hand, will focus light from a given star to a very tiny spot, effectively a geometric point, and there is no geometrical aberration in this image.

The Scottish mathematician James Gregory, whom we met in chapter 1, invented the first reflecting telescope in 1661. Gregory had studied the mathematical properties of conic sections, and he noticed that if conics were used as optical reflectors, the resulting images could be freed of spherical aberration. The Gregorian telescope (Fig. 8.8) used a large concave parabolic mirror to collect light and reflect it to a concave elliptical mirror, which in turn focused the light to a point behind a hole in the first mirror. Here, the observer viewed the image through a small eyepiece.

Although a creative and mathematically elegant design, in practice the Gregorian reflector still suffered from limitations in the grinding tolerances

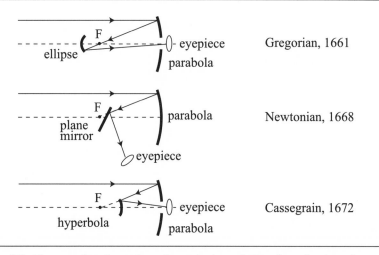

Figure 8.8. The use of conic-section mirrors in three designs for reflecting telescopes.

attainable in the seventeenth century. We now know that a mirror's surface must be geometrically precise to within about ±0.25 thousandths of a millimeter to permit it to approach the theoretical limit of its resolving power. It turns out that a parabolic surface can be ground close to this tolerance by hand, while a precise ellipsoidal surface is considerably more difficult to create. Worse (although this would not be proved until more than century later by Jesse Ramsden), if an optical system contains two sequential concave reflectors, regardless of their shapes, the combined effect is to magnify any geometrical imperfections in either surface. For these reasons, the actual performance of the Gregorian telescope was somewhat disappointing.

Just a few years later, in 1668, Isaac Newton (1642–1727) invented a slightly different telescope that also used a parabolic reflector (see Fig. 8.8). In the Newtonian design, the secondary reflector was a flat mirror, easily built to fine tolerances, and the image was projected from the side of the telescope tube. Newton demonstrated his second prototype of this telescope to the Royal Society of London early in 1672, and the members were so impressed that they elected him a fellow at that same meeting.[6] The economy of Newton's design makes it popular among amateur astronomers even today.

Later that same year, Guillaume Cassegrain published his own design for a reflecting telescope (see Fig. 8.8). Cassegrain also used a concave parabolic reflector as the primary mirror, but his secondary mirror was convex, in the shape of a hyperbola that shares a focus with the parabola. As in the Gregorian design, the image was focused through a hole in the primary mirror. Newton argued vehemently that Cassegrain's instrument was no improvement on his own design, and Cassegrain retreated into obscurity in the face of this criticism. Only much later would it be shown that Cassegrain's combination of a concave parabolic mirror and a convex hyperbolic mirror worked to cancel geometric imperfections in both surfaces. Today, numerous telescopes at major observatories make use of Cassegrain's design.

The seventeenth-century flurry of activity in the design of optical systems was the direct result of the new analytic geometry of Descartes and Fermat. No longer did one need to guess about the optical effects of a particular curved reflector, for one now could analyze such effects algebraically before grinding the surfaces and building an instrument. Analytic geometry explained why spheres are inferior surfaces for focusing clear images, and why parabolic reflectors are much better. This mathematical system also provided the computational tools to predict how elliptical and hyperbolic reflectors might be used in creative combination with parabolic reflectors. With the advent of analytic geometry, the conic sections had become curves not just to be sketched or studied in the abstract, but to be physically constructed to serve practical needs.

Motion in Gravitational Fields

By the mid-seventeenth century, most European scientists (then called "natural philosophers") were familiar with Kepler's laws of planetary motion. The requisite mathematics had a firm footing in analytic geometry, and computations could predict the position of any planet for any future date of interest. Predicting, however, is not the same as explaining. Kepler's laws did not address *why* the planets move in ellipses as they do, and mysteries remained about how these laws might apply to nonplanetary objects like moons, comets, meteors, or falling apples.

A late outbreak of bubonic plague struck England in 1666, and the administrators of Cambridge University wisely closed that institution's doors for a year. Isaac Newton, then a twenty-four-year-old student, went home to his family's farm and began thinking about the structure of the universe. Maybe he noticed an apple falling from a tree, and maybe he didn't. No matter, for since earliest times it had been no secret that things fall—whether apples from trees, bison stampeding over a cliff, or arrows in flight. What puzzled Newton was the moon that hovered overhead. Why doesn't the moon fall, just like everything else that has no visible means of support?

There seemed to be but two feasible explanations. One possibility was that the laws of physics vary from one place to another in space. Alternatively, maybe the moon *does* fall continuously, yet it never gets any closer to Earth because of the way it moves tangentially as it falls.

Newton rejected the first notion. If the laws of nature differed from place to place, we would observe total chaos in the universe. As Earth speeds through the solar system, it sometimes occupies space where the moon has been just a few hours earlier. If objects don't fall in the pocket of space occupied by the moon, then we ought to expect to wake up some mornings to find ourselves floating on the ceiling. That this doesn't happen, and that laboratory experiments can generate replicatable results from month to month, suggests that the laws of nature are universal: consistent both in time and space. Newton accordingly developed a conviction that nature is not capricious, and that whatever drives an apple to fall must also explain why the moon seems to hover.

That left the second alternative: maybe the moon *is* falling. Newton devised an imaginary experiment. Suppose we put a cannon on top of a tall mountain, and we shoot a series of cannonballs parallel to Earth's surface, as shown in Figure 8.9. The first shot has only a tiny charge of explosive,

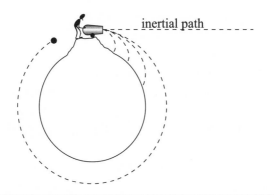

Figure 8.9. Newton's imaginary cannon experiment. All of the cannonballs fall, regardless of their tangential speed. If they did not fall, they would follow a straight-line inertial path through space.

and the cannonball barely makes it out of the muzzle before falling to the ground, just like an apple falling from a tree. The second shot is propelled by a larger charge, and the projectile now follows a parabolic arc as it falls. The next shots, fired with ever-increasing amounts of propellant, eventually disappear over the horizon as they fall. Finally, with enough gunpowder, a speeding cannonball ought to completely circle Earth without hitting the surface as it falls. This, reasoned Newton, is what the moon does as it orbits Earth. The moon, therefore, *is* falling; it just never strikes Earth's surface because it is moving fast enough to keep from getting any closer as it falls. If the moon or the cannonball did not fall, these objects would follow straight-line paths through space. Newton referred to such a "natural" straight-line path as the *inertial path*. Anyone who has ever failed to negotiate a turn while driving on an icy road is familiar with this natural tendency of objects to move in a straight-line path at a constant speed.

Moving objects deviate from inertial paths only when forces act to deflect their motions. For a car rounding a bend, the deflecting force is lateral friction between the tires and the road. For the moon orbiting Earth in space, the deflecting force must be gravity—the same mysterious force that causes a cannonball to fall. Clearly, gravity depends on the mass of the central attracting object, and it is also affected by the separation between objects.

But what mathematical law does gravity conform to? Newton began by considering the special case of a circular (rather than elliptical) orbit. No object can move in a circle unless it experiences a deflecting force equal to mv^2/R, where m is the object's mass (in kilograms, for instance), v is its speed, and R is the radius of the circular path. The force specified by this equation is usually referred to as the *centripetal* force. This formula is easily verified in the laboratory, and Newton reasoned that it ought to be universal in its application. He then postulated various equations that might describe the force of gravity, equated these expressions to the quantity mv^2/R, and worked out their mathematical consequences. The postulate that led to the most interesting result was this:

$$\text{force of gravity} = G\frac{mM}{R^2},$$

where m is the orbiting mass, M is the central mass, R is the radial separation between the two masses, and G is a universal gravitational constant.

Equating this expression to the centripetal-force formula gives

$$\frac{GmM}{R^2} = \frac{mv^2}{R},$$

and the orbiting mass m cancels out. The physical implication is that an object's path through space does not depend on its own mass, but rather on the central mass that is attracting it gravitationally. This alone is a profound conclusion, for it explains how both Earth and our moon manage to orbit the sun once a year at the same average radius, despite the fact that the moon's mass is considerably less than the mass of planet Earth.

Going further, we can substitute for the orbital velocity $v = 2\pi R/T$, where T is the orbital period (the time for one complete orbit) and $2\pi R$ is the orbit's circumference. After some rearranging, this leads to the conclusion that if the force of gravity is GmM/R^2, then

$$\frac{R^3}{T^2} = \frac{GM}{4\pi^2}.$$

This result struck Newton right between the eyes, for it is Kepler's third law of planetary motion, plus something more. Not only is the ratio R^3/T^2 equal to a constant, as Kepler had discovered earlier from Tycho Brahe's data, but now that constant has been related to the central mass, the value of π, and a universal constant G that describes the strength of the gravitational force everywhere in the universe.

Still, this didn't *prove* to Newton that he had found the correct mathematical form for the law of gravity, for the above analysis is based only on the special case of a circular orbit. Could this same gravitational-force formula also account for Kepler's elliptical orbits? Here, the problem is considerably more complicated, and neither algebra nor analytic geometry suffices. To solve the general problem of planetary orbits, Newton needed to develop a new branch of mathematics, which today is referred to as *calculus*.[7] This mathematical system allows one to go beyond linking simple variables, and to explore the rates at which the variables change with respect to one another. The resulting equations are called *differential* equations, and their solutions are not mere numbers, but families of mathematical functions. In the planetary-orbit problem, the radial distance from the sun is a variable, the planetary velocity is a variable, and the rate of change of velocity (the acceleration) is also a variable.

In Cartesian coordinates, the differential equations for motion under the influence of gravity are quite messy. Newton noticed, however, that the problem assumes a simpler form when recast in polar coordinates (r, θ), with the central mass centered at the origin of the coordinate system. The two resulting equations are

$$2\left(\frac{dr}{dt}\right)\left(\frac{d\theta}{dt}\right) = r\frac{d^2\theta}{dt^2} = 0$$

and

$$\frac{d^2r}{dt^2} - r\left(\frac{d\theta}{dt}\right)^2 = -\frac{GM}{r^2}.$$

Here, the notation dr/dt represents the instantaneous rate at which the radial coordinate r varies in time (the derivative of r with respect to t), $d\theta/dt$ is the corresponding rate at which the angular coordinate θ varies in time, and the quantities d^2r/dt^2 and $d^2\theta/dt^2$ are the rates at which the previous rates change in time (the second derivatives).

Readers who have studied higher mathematics but have never tackled this particular problem will gain considerable insight into Newton's achievement by trying to solve the above equations on their own. It is far from an easy task, and in fact few mathematics textbooks treat systems of coupled differential equations like these. Yet Newton, while still in his mid-twenties, not only formulated this problem but also developed the mathematics needed to solve it.

The solution to the first equation is

$$r^2\frac{d\theta}{dt} = \text{a numerical constant.}$$

This says that when an object is far from the attracting mass (i.e., r is large), the angular position changes slowly, whereas when it is close to the attracting mass, the angular position changes quickly. Both verbally and numerically, this is equivalent to Kepler's second law: the line between the sun and a planet sweeps through equal areas in equal time intervals.

The second equation has the solution

$$r = r_o\left(\frac{1 + e}{1 + e\cos\theta}\right),$$

where r_0 is the radial position of the object along a reference line $\theta = 0$, and e is a constant of the motion, which is inversely proportional to the product MG. Newton must have initially stared in disbelief at this result, for this is the general equation, in polar coordinates, for *all the conic sections*! The constant e is none other than the eccentricity.

With these solutions, Newton demonstrated that all three of Kepler's laws of planetary motion were consequences of a single universal law of gravity. In the sun's gravitational field, the planets have no choice but to orbit in circles or ellipses, and to travel faster when closer to the sun and slower when farther away. Further, the masses of the planets themselves have no influence on their own motions; the only mass that affects an orbit is the central mass, the sun. And if a comet should appear, these same solutions render its motion every bit as predictable as the motions of the planets.

Newton's mathematical conclusions went further still, for they established that the parabola and the hyperbola are also possible paths. The parabolic trajectories of cannonballs on Earth are, in this mathematical system, but a special case of the more general conic-section solution. Meanwhile, nonperiodic comets are indeed observed to follow hyperbolic paths through space. Equally important, the motions of planetary moons could now be predicted simply by replacing the sun's mass in the law of gravity with the central planet's mass. Before long, a whole host of other phenomena yielded to explanation in terms of the law of gravity: for example, the tides, the fact that the moon always presents the same face to Earth, the ring systems of the outer planets, and the observation that large planets are more spherical than small moons and asteroids.

The Many-Body Problem

The successes of Newton's theories of gravity and motion led many to conclude that the physical universe is structured on strictly mathematical principles, and that once the Creator started the mathematical clockwork, all subsequent events were programmed to unfold in an inevitable sequence. In this view, referred to as *determinism*, the universe has but one possible future. Accordingly, given the proper mathematical understanding, it should be possible, at least in principle, for humans to predict every event that is yet to happen.

In fact, Newton's mathematical science was quickly applied in ways he'd never dreamed of, successfully predicting the behavior of engines, machinery, and structures before they left the drafting table. On the other hand, natural phenomena like weather, earthquakes, and volcanic eruptions have to this day remained stubbornly intransigent to Newtonian determinism. Had Newton begun his inquiry into the mysteries of the universe by studying earthquakes, or the epidemic that closed the doors of his university in 1666, he never would have made as much progress as he did. His good fortune was to have chosen a physical system where the variables were few in number, and which behaved deterministically—at least almost.

Of all the mass in our solar system, the sun accounts for 99.8%. The planets are tiny in comparison and widely spaced. Newton's analysis assumed that each planet could be treated as an object unto itself, influenced *only* by the sun's gravity. What he solved, then, was a generalizable two-body problem. In any force field having a $1/r^2$ dependence, the two-body problem always results in orbits that are conic sections.

Yet our solar system is not really a two-body system. Planets *do* interact with each other gravitationally, in a subtle yet complicated manner that depends on their varying distances from one another. As a result, the orbits of planets deviate slightly from perfect conic sections. Just as we saw earlier that there is no perfect circle in nature, we are forced to conclude that there are no perfect conic sections either, even in space.

If not perfect conics, then what curves do describe the planetary orbits? Newton himself attempted to answer this question by seeking a general solution to the three-body problem (two planets and the sun), but he found the problem to be intractable. Since then, it's been established that no stable closed-form solution exists when three or more bodies interact through gravity. Modern computers can number-crunch solutions to special cases of these many-body problems, but the results are never describable in terms of analytic mathematical functions. In such calculations, the planetary orbits never quite close on themselves, and as one planet passes near another, both of them deviate from their ideal elliptical orbits.

Computer calculations make it clear that we are very lucky to live on a planet that orbits just one star, our sun. In the broader universe, at least half the stars are part of multiple-star systems, in which two or more stars orbit each other. Toss a few planets into such a system, and it is impossible for them to retrace their orbital paths as the suns themselves move around. If our Earth were in such a system, it would cavort about in an irregular nonperiodic orbit, and might even encounter a set of conditions where it was expelled from the system into interstellar space. Clearly, life could never evolve under such unstable conditions.

In 1781, the astronomer William Herschel discovered a new planet beyond the orbit of Saturn, which he named Uranus (the mythological father of Saturn). After observing this planet for about twenty years, astronomers noticed that its orbit deviated slightly from Kepler's laws and the predictions of Newtonian dynamics. Worse, these orbital variations could not be explained in terms of gravitational interactions with Saturn and Jupiter, for Uranus sometimes wobbled when it was nowhere near these other planets. This was troublesome, for it suggested that Newton's laws might not be universal after all, but rather that they might begin to fail at large distances from the sun.

There was, however, an alternative explanation for the irregularities in the motion of Uranus: perhaps there was yet another undiscovered planet even farther from the sun that was influencing the orbit. Such a distant planet would be very dimly lit, and virtually impossible to find telescopically unless one had a very good idea of where to look for it. And so, in the 1840s, a number of mathematician-astronomers attempted to compute the orbit of a hypothetical new planet whose gravitational influence could account for the observed motion of Uranus. This required solving a three-body problem where nothing at all was known about the outermost body, and the computations proceeded by laborious approximation and trial and error. In 1845, after two years of calculations, the astronomer J. C. Adams announced a position and orbital period for the hypothetical planet. British astronomers dutifully looked for it, failed to realize that they'd actually charted it, then gave up the search. Soon after, the French mathematician Leverrier independently completed his own calculations and interested two German astronomers in searching for the planet. Within a half-hour of starting their search on September 12, 1846, they discovered the planet Neptune.

This discovery was a wonderful validation of the explanatory power of Newton's law of gravity. When later observation revealed that Neptune's orbit also had small irregularities, the natural thing to do was to search for a ninth planet. This search began in 1905, and in 1930 it resulted in the discovery of Pluto. Since then, other tiny discrepancies in planetary orbits have enabled astronomers to discover a large number of other bodies in space, including a small object orbiting the sun beyond Pluto and an icy body (Chiron) orbiting between Saturn and Uranus. The law of gravity is also the main tool in the search for planets orbiting other stars; tiny wobbles in some stellar motions seem to be evidence of the gravitational effects of orbiting objects that cannot be, or have not been, imaged directly. At present, several dozen such distant solar systems have been discovered in this manner, and crude images have been obtained for a few of the closest.

Although the general many-body problem yields but few stable solutions, our solar system as we know it *is* relatively stable, its planetary motions deviating only slightly from true ellipses that have small eccentricities. The early solar system, however, swarmed with a beehive of moving objects, with wildly varying intersecting orbits. How did the present relative stability evolve from our solar system's earlier chaos?

Again, an answer can be found in the law of gravity, aided by the computational power of modern computers. The early solar system contained many millions of objects of different sizes, moving almost at random except that they interacted gravitationally. Although we don't know much about the sizes or orbits of these early objects, we can simulate thousands of likely scenarios. In every case, such computer simulations show that collisions initially were frequent, and that objects with high-eccentricity orbits, on an average, collided much more often than objects with nearly circular orbits. Eventually, under any reasonable set of initial conditions, most of the high-eccentricity orbits are found to disappear, and the low-eccentricity (nearly circular) orbits account for an increasing portion of the swarming mass. Looking at the process in more detail, we find that collisions have one of two results: (1) one or both colliding objects fracture into smaller pieces, or (2) the larger body absorbs the mass of the smaller body, its orbit altered by the dynamics of the impact. Near misses also alter the orbits of the interacting bodies, with the least-massive body experiencing the greatest change in its orbit. When an orbit is altered so its eccentricity exceeds unity, that object leaves the solar system forever. The objects that remain behind are the planets, their satellites, and their rings: objects with nearly circular orbital geometries, for which collisions are unlikely.

Although the many-body problem cannot be solved exactly, the statistics of numerical solutions suggest that, given enough time, the chaos of any early single-star solar system will eventually clear out its own detritus and settle down to motions that can reasonably be approximated by the conic sections. In this sense, ancients like Ptolemy were right in assuming that the universe prefers circular orbits, for it is the low-eccentricity orbits that indeed are the most stable. And, given such relative stability, it becomes possible to use any observed deviations from circles and ellipses to draw mathematical inferences about the masses and orbits of other interacting bodies. In a universe of perfect circles, our prospects for gaining additional knowledge about small distant objects would be very limited. We learn, not from circles themselves, but rather by exploring why the circles and other conic sections we observe are less than perfect.

CHAPTER 9

OSCILLATIONS

As we saw in chapter 3, our concept of time is intimately bound up with the circle and cyclic phenomena. The motions of gears and clock hands, the swinging of a pendulum, the vibration of a quartz crystal, and the electromagnetic radiation in atomic clocks all draw their mathematical descriptions from circular geometries.

Most scientists prefer to avoid philosophical arguments about what time *is*; instead, they've reached consensus on how time is to be *measured*. Whether we talk about milliseconds or millennia, our operational definitions are that time is what we measure with clocks, and that a clock is a device that counts the cycles of some (nearly) invariant natural cyclic process. Today, one second is internationally defined as the duration of 9,192,631,770 periods of the radiation corresponding to the transition between two specific hyperfine levels of the ground state of the cesium-133 atom.[1] With the second thus defined, the day, month, and tropical year follow.

In centuries past, of course, time was defined by the astronomical rather than the atomic, and the challenge in building a clock was to divide astronomical cycles rather than to multiply atomic cycles. We've seen how Earth's natural celestial clock led early humans to count days, months, and years. For shorter periods, say hours or quarter-hours, shadow-clocks (sundials) gave better precision than direct visual observations. If, however, the sun did not cooperate, or if someone needed to measure times shorter than around a quarter-hour, a different approach was needed.

Time candles and graduated oil lamps were early alternatives. Although the principle is simple (the longer a flame burns, the more fuel it consumes), a rate of fuel consumption unfortunately depends not just on time, but on additional factors such as: variations in the chemistry of the fuel, the temperature of the fuel, and the air temperature. Because there is nothing cyclic taking place in the combustion process, such devices are difficult to regulate. As a result, the accuracy of flame clocks was never very impressive.

A more successful early approach was the clepsydra, or water clock, invented independently in ancient Egypt and China.[2] Originally consisting of little more than water dripping from a small hole near the bottom of a funnel-shaped vessel, this device went through a series of refinements over the centuries. Several water clocks dating from c. 300 B.C.E. indicated time on a dial, which was rotated by a lever attached to a float in the water vessel. Still later devices kept the water level constant in an upper vessel, from which the water dripped into a lower vessel at a steady rate. None of these devices, however, was very accurate by today's standards. The major problem was that the viscosity of water (or any other fluid, for that matter) is temperature dependent. If a simple water clock is calibrated by placing it next to a sundial during the day, it will run slower during the cooler night that follows, and it will be out of step with the sundial the next day.

In 725 C.E., a Chinese military official, I'Hsing, built a more sophisticated clepsydra in which the dripping water drove a water wheel that rotated at a constant rate regardless of temperature. The wheel, in turn, was coupled to a dial that indicated the passage of time. In 1090, the Chinese scientist Su Sung elaborated on this design and constructed a huge astronomical clock standing around forty feet tall, with tiers of puppets that paraded about to announce the hours and quarters. Unfortunately, detailed documentation of the workings of these devices no longer exists, and we can only speculate on the principle that was used to maintain the wheel's constant rate of rotation. Possibly a flyweight-controlled mechanism was used, similar to the governors on some European windmills of the late Middle Ages. It seems unlikely that these early Chinese clepsydras were regulated by a mechanical oscillator, for if they were, such a superior design surely would have persisted into the following millennium.

In medieval Europe, simple clepsydras were commonly used in monasteries to alert the sacristan to his duty of ringing a tower bell. Around the last quarter of the twelfth century, the purely mechanical clock appeared, powered by a hanging weight that rotated a gear that drove the mechanism. Alone, however, the rotation of a gear does not make a clock; some periodic phenomenon that regulates the gear's rate of rotation is necessary. In the simple clepsydra, the periodic phenomenon was dripping water. In the first mechanical clocks, it was the reversing motion of a horizontal bar pivoting at its center (called a foliot), which was kicked back and forth by the action of a gear's teeth against a lever (called a verge).

All mechanical or electronic clocks need three functional elements: a source of energy, an oscillating system, and an escapement. The escapement is a device that feeds energy to the oscillator in small increments, just sufficient to keep the oscillating system in motion. The motion of the oscillator is coupled to an indicator that displays the time (moving clock hands, for instance).

The verge-and-foliot oscillator, alas, was no more regular in its periodicity than dripping water; it was just less messy. When, in the fourteenth century, large tower clocks of this design began appearing throughout Europe, they typically remained in step with a sundial only to within about a half-hour per day. The problem with the verge-and-foliot mechanism was that its cyclic rate depended to a great extent on the friction in the bearings, which in turn depended on temperature and other factors. More accurate timekeeping would need to wait until other cyclic phenomena were found whose periods were not affected by inevitable dissipative mechanisms such as friction.

More holistically, the problem of timekeeping may be viewed this way: Nature presents us with cyclic events in the heavens. We wish to divide the periods of these natural cycles into smaller units, and to do so with high accuracy. Given that we cannot couple a gear train directly to the heavens, we are confronted with the problem of assuring that two quite different cyclic systems (astronomical motion and our Earth-bound clock) can be brought into synchronicity and remain synchronized. It is amazing that this can be done at all, for what connection does dripping water (for instance) have with Earth's motion around the sun? The only connection is that both systems are subject to natural laws that are consistent across the fabric of space and time. Successful mechanical clocks stand as wonderful confirmations of Pythagoras's earlier speculation that nature is orderly enough to be mathematically predictable.

The Pendulum

Even today, however, we have no precise theory of friction. Friction arises from interactions at the molecular level, and a host of variables can influence the amount of friction developed between a given pair of surfaces.[3] On Earth, almost all motions are affected by friction, whether it be tires on a road, an aircraft speeding through the air, or a swinging pendulum. Only in the near vacuum of space do moving objects experience virtually no friction (although gravitational interactions can still induce tidal friction, which over the long term can affect the periodic motions of planets and their moons).

Given that friction affects all terrestrial motions, it may seem that there could be little hope of ever building an accurate mechanical clock. A clock needs an oscillating device to regulate it, but friction renders irregular the amplitude of any mechanical oscillator. Doing a simple computation gives a sense of the seriousness of this problem. If, for instance, we base a clock on an oscillator whose period is 1.000 s ± 0.003 s, seemingly a fairly accurate figure, over the course of a day this clock is still capable of gaining or losing a full 5 minutes!

In 1583, during a church service at the cathedral in Pisa, the nineteen-year-old Galileo Galilei watched with curiosity as a set of hanging lamps swung back and forth, stirred into motion by air currents. Although the amplitudes of the swings were all different, the lamps nevertheless stayed in step as they swung. Using his pulse to time the periods of the motions, Galileo found, to his surprise, that as a swing lost amplitude due to friction with the surrounding air, it did not "slow down" in the sense of aquiring a longer period. Later, he verified this discovery of the isochronism (constant period) of the pendulum in more carefully controlled experiments. As shown in Figure 9.1, the period of a pendulum is relatively unaffected by the amplitude of its oscillation; the only way to change the period is to alter the pendulum's length.

By the early 1600s, astronomers were using pendulums to time astronomical events such as eclipses. Galileo struggled for many years to invent a complete mechanical clock that would be regulated by a pendulum, and in fact he did solve this problem just before he died, when he was totally blind. His son later commissioned such a clock to be built, but he too died before the device was completed.

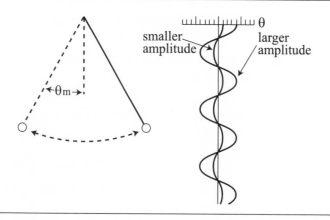

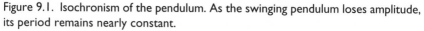

Figure 9.1. Isochronism of the pendulum. As the swinging pendulum loses amplitude, its period remains nearly constant.

The first practical pendulum clock was apparently built by the Dutch scientist Christiaan Huygens sometime around 1657.[4] While inventing this instrument, however, Huygens discovered that Galileo had been wrong about the isochronism of the pendulum. It turns out that a pendulum's period is independent of amplitude only for small amplitudes; at larger amplitudes, the period does vary slightly with changes in amplitude (Table 9.1). Galileo had simply been unable to measure time periods precisely enough to notice this effect. Without an accurate clock already in hand, how could he?

Simple Harmonic Oscillations

Vibrations are ubiquitous in nature. Tree branches sway in the wind, ships bob and pitch in the seas, and earthquakes shake buildings. Such natural vibrations can be complicated and erratic. In many cases, however, the vibrating object periodically repeats its path in a simpler motion referred to as an *oscillation*.

After Newton formulated his laws of mechanics in the late 1600s, it became possible to analyze oscillating systems mathematically, and to predict an oscillator's period without actually measuring it. Oscillations arise only in systems that are near a state of *stable* equilibrium, where any disturbance gives rise to forces that restore the initial state. When a moving car hits a bump, for instance, the body bounces a few times as the springs bring it back to its original level. Such oscillations do not arise with *unstable*

Table 9.1. Period of a simple pendulum 1.0000 meter in length, as a function of amplitude. The oscillations are simple harmonic (and therefore isochronous) only for very small amplitudes.

Amplitude (°)	Period (s)
~0	2.0064
2	2.0065
4	2.0071
6	2.0078
8	2.0088
10	2.0102
15	2.0151
20	2.0217
30	2.0413
45	2.0866
60	2.1533
75	2.2453
90	2.3683

Note:
  Periods can be calculated from $T = 4\sqrt{(L/g)}K(k)$, where L is the length of the pendulum, g is the gravitational acceleration (9.80665 m/s^2), and $K(k)$ is a function called the complete elliptic integral of the first kind. The argument of this function, k, is related to the pendulum's amplitude θ_o by $k = \sin(\theta_o/2)$. See G. R. Fowles, *Analytical Mechanics* (New York: Holt, 1970), pp. 105–106.

systems—for example, a coin balanced on its edge. If such a coin is disturbed, it topples away from its equilibrium state, and that's that.

Newton's force laws tell us that an object's acceleration is proportional to the net force acting on it. If an object in stable equilibrium experiences a linear restoring force when displaced in one dimension, that object's motion satisfies a differential equation of the form

$$\frac{d^2x}{dt^2} = -Cx.$$

The constant C is related to the physical parameters of the particular system: its mass, buoyancy, and/or elasticity, for example. The solution to this differential equation is a family of curves that has the mathematical form

$$x = x_m \cos(\omega t + \phi_o),$$

where ϕ_o is the initial phase angle (usually taken to be zero), x_m is the amplitude (the maximum displacement from equilibrium), and ω represents the angular frequency of the oscillation:

$$\omega = \frac{2\pi}{T} = \sqrt{C}.$$

Here, T is the period of the physical oscillation, and C is the constant in the original differential equation.

And what does this all mean? For one thing, it means that the physical laws Newton developed to account for the motions of planets apply equally well to other moving bodies, even small bodies executing oscillatory motion on Earth rather than orbital motion in space. Although celestial bodies are not physically connected with our mechanical clocks, motions in the heavens can nevertheless be subdivided by clocks on Earth because both sets of motions follow the very same mathematical laws.

More specifically, for objects subject to linear restoring forces, Newton's laws predict that the period of oscillation will be independent of the amplitude of oscillation. Such oscillations are said to be *simple harmonic,* and their time-dependent motion can be described by the simple cosine curve we saw in Figure 9.1.

Instances of simple harmonic oscillations abound; a few common examples are listed in Table 9.2. Disturb a mass hanging on a spring, or charge an electrical capacitor whose terminals are connected to an inductor, and the system oscillates with the period given by the equation in this

Table 9.2. Periods of some simple harmonic oscillators.

Oscillating system	Period	Independent variables
Simple pendulum (small amplitudes)	$T = 2\pi\sqrt{\dfrac{L}{g}}$	L = length g = gravitational acceleration
Mass on spring	$T = 2\pi\sqrt{\dfrac{m}{k}}$	m = mass k = spring stiffness
Balance wheel on torsion spring	$T = 2\pi\sqrt{\dfrac{I}{\kappa}}$	I = moment of inertia κ = torsional stiffness
Electrical circuit	$T = 2\pi\sqrt{LC}$	L = inductance C = capacitance

table, independent of its amplitude. The mathematical analysis also tells us that, unless we use a measurement unit that has been defined as some multiple of π, it is impossible for the period of an oscillator to assume an integer or other rational value. Pi is always a factor in the period of an oscillator.

Pi, in turn, suggests that a circular geometry is somehow connected with the phenomenon of oscillation. Let's explore this connection further, by considering a bolt fastened to the rim of a wheel of radius R. Imagine this wheel spinning at a constant rate, so the bolt completes a revolution of 2π radians in the time period T. Then, as shown in Figure 9.2, the angle θ increases linearly in time: $\theta = (2\pi/T)t$. The horizontal displacement x of the orbiting bolt is $x = R \cos \theta$, where x has a maximum value of R when $\theta = 0$. Combining these relations leads to

$$x = x_m \cos\left[\frac{2\pi t}{T}\right] = x_m \cos(\omega t),$$

which is identical to our earlier solution for the motion of a simple harmonic oscillator in one dimension! Simple harmonic oscillations, in other words, may be viewed as the projection of uniform circular motion onto a straight line. It would be possible, for instance, to set up an oscillating mass-spring system below the spinning wheel in such a way that the

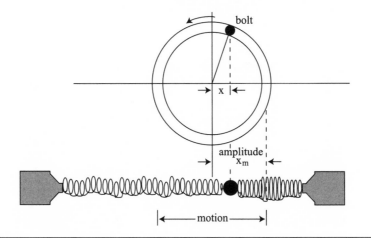

Figure 9.2. A simple harmonic oscillation is the same as the projection of a uniform circular motion onto a line. For both the bolt on the spinning wheel and the mass-spring system, $x = x_m \cos(2\pi t/T)$.

shadow of the revolving bolt falls on the oscillating mass and follows the mass perfectly during its sinusoidal motion.

In fact, such a projection arises physically during the generation of alternating-current electricity. In an electrical generator, a coil of wire is rotated at a constant angular speed in a fixed magnetic field. When the coil slices through the magnetic-field lines at a right angle, the induced electric current is a maximum; 90° later, when the coil travels parallel to the field, the electric current is a minimum. The resulting alternating current oscillates as a cosine curve (or a sine curve, which is the same function with the starting point shifted 90°). The electricity that enters our homes is a simple harmonic oscillation. As a result, the period ($\frac{1}{60}$ second, in the United States) remains very nearly constant even though the peak current and voltage fluctuate significantly with changing loading conditions.

Oscillations and Fourier Series

For those of us who value order higher than chaos, it's always deeply re-assuring to discover that a natural phenomenon conforms to a set of math-ematical rules. Nature, however, is by no means compelled to respect any particular line of human mathematical reasoning. In fact, many oscillations in the physical world are *not* simple projections of uniform circular mo-tion, but rather deviate to some extent from this ideal. One common ex-ample, shown in Figure 9.3, is the motion of a reciprocating piston that is

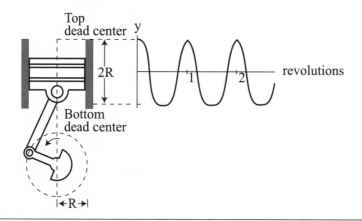

Figure 9.3. The motion of a reciprocating piston is anharmonic if it is connected to a crankshaft that rotates at a constant speed. Most of the piston movement takes place during the top half-circle of crankshaft rotation.

linked to a rotating crankshaft. This mechanism is found in piston engines and pumps.

Let's picture such a crankshaft as it rotates at a constant rate. Clearly, the piston travels a distance of $\pm R$ about its center position, where R is the radius of the crank circle. Equally clearly, both the oscillation of the piston and the rotation of the crankshaft have the same period T. Note, however, that as the crank rotates through the top 180° of its circle (in a time of $T/2$), the piston moves through *more than half* of its total stroke ($>R/2$). This excess motion arises because the connecting rod does not move colinearly with the piston motion, but rather swings outward as the crank rotates away from top dead center. In the following bottom 180° of rotation, the piston moves through *less than half* of its total stroke, as the connecting rod swings in and out around bottom dead center. Depending on the length of the connecting rod relative to the radius of the crank circle, an engine's piston may execute as much as 60% to 70% of its total stroke during the half-period around top dead center. Note from Figure 9.3 that this piston's motion clearly deviates from a perfect cosine curve. Mechanical engineers must consider this anharmonic effect when timing the opening and closing of an engine's intake and exhaust valves.

Motions like that of the piston appear to be considerably more complicated than simple harmonic motions. Around 1807, however, the French mathematician and physicist Jean Baptiste Joseph Fourier (1768–1830) made a wonderful discovery: *all* periodic motion, even anharmonic oscillations, can be reduced to sine and cosine functions.[5] The trick is to view an anharmonic oscillation as a *sum* of a set of simple harmonic oscillations of decreasing amplitudes and periods. Mathematically, a Fourier series contains an infinite number of terms, each term a sine or cosine function. In practice, limited by finite precision of measurement, scientists and engineers can truncate a Fourier series and ignore those higher harmonic components whose amplitudes are too small to measure.

An example of a Fourier series is shown graphically in Figure 9.4. Here, we begin with a fundamental sinusoidal motion of amplitude 1 unit and period T. To this we add a second harmonic sinusoidal motion of amplitude $\frac{1}{2}$ and period $T/2$. The third harmonic has amplitude $\frac{1}{3}$ and period $T/3$, and so on. Summing these harmonic motions results in a sawtooth-shaped oscillation of amplitude $\pi/2$ and the fundamental period T. Although the resulting motion is clearly not sinusoidal, we see that it is indeed made up of superimposed sinusoidal components.

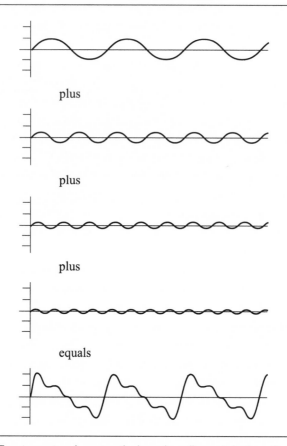

Figure 9.4. A Fourier series. A sawtooth-shaped oscillation is built up from the super-position of the simple sine waves shown.

The problem also arises in reverse: Given a specific nonsinusoidal oscillation, what are its harmonic components? The mathematical procedure for answering this question is referred to as *Fourier analysis.*[6] In Figure 9.5, for instance, we see a triangular-shaped oscillation. A Fourier analysis of this periodic function reveals that it is composed of a series of cosine functions whose amplitudes decrease in the proportions $\frac{1}{1}^2$, $\frac{1}{3}^2$, $\frac{1}{5}^2$, and so on, and whose periods decrease in the proportions $\frac{1}{1}$, $\frac{1}{3}$, $\frac{1}{5}$, . . .

Results of such Fourier analyses are of more than abstract interest, for it turns out that the leading terms of a Fourier series can actually be observed in a variety of natural phenomena. A frequency measurement of a triangular electrical oscillation, for instance, reveals that it has not just one frequency, but rather a whole set of frequencies in odd-integer multiples.

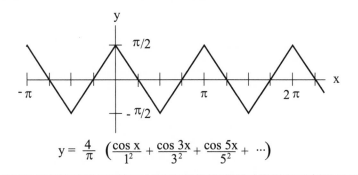

$$y = \frac{4}{\pi}\left(\frac{\cos x}{1^2} + \frac{\cos 3x}{3^2} + \frac{\cos 5x}{5^2} + \cdots\right)$$

Figure 9.5. Fourier analysis of a triangular wave. Every periodic waveshape can be resolved into its sine and cosine components.

If such a triangular electrical signal enters a circuit that is tuned to respond to only the third harmonic frequency (for instance), only a cosine wave of that particular frequency will pass through. Similarly, an earthquake wave with a 2-second period may be capable of inducing vibrations in a building whose natural period of oscillation is only 1 second. In this instance, it is the second harmonic of the earthquake waveform that sets up a 1-second resonant vibration in the structure. Physical systems, it turns out, actually respond to each of the mathematical terms in a Fourier series just as if these individual harmonic oscillations were physically present.[7]

Fourier Series and Circles

We can gain further insight into the physical meaning of a Fourier series through a geometrical diagram. We've seen that every simple harmonic oscillation can be viewed as the projection of a uniform circular motion onto a straight line. We should expect, therefore, that a more complicated oscillation can be represented as the projection of a *series* of circles onto a line.

Figure 9.6 shows four such circles whose radii, relative to the largest circle, are in the ratios $\frac{1}{1}$, $\frac{1}{3}$, $\frac{1}{5}$, and $\frac{1}{7}$. As the center of each circle moves counterclockwise along the circumference of the next larger circle, we adjust the periods of their rotations to reflect the same set of ratios $\frac{1}{1}$, $\frac{1}{3}$, $\frac{1}{5}$, and $\frac{1}{7}$ (the smaller circles rotate faster). All four of these circles begin in the position where they are stacked up vertically, and our diagram represents a snapshot of the situation at a later time.

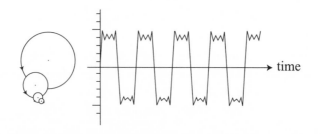

Figure 9.6. A Fourier series can be interpreted geometrically as the projection of a system of superimposed circular motions.

We now follow circle #4 as it rides upon the nest of spinning circles beneath it, and we project the motion of a point on its circumference onto a line. A trace of this projection is shown in the time graph in Figure 9.6. The result is a periodic but distinctly nonsinusoidal motion, which can be described mathematically by

$$y(t) = (1)\sin\left(\frac{2\pi t}{T}\right) + \left(\frac{1}{3}\right)\sin\left(\frac{3 \cdot 2\pi t}{T}\right) + \left(\frac{1}{5}\right)\sin\left(\frac{5 \cdot 2\pi t}{T}\right) + \left(\frac{1}{7}\right)\sin\left(\frac{7 \cdot 2\pi t}{T}\right),$$

where T is the fundamental period of oscillation, equal to the time for one rotation of the largest circle. This is a Fourier series that has been truncated after the fourth term. If we included more terms, (i.e., more circles turning upon circles), the resulting graph would more nearly approximate a series of alternating horizontal line segments.

The curious thing is this: Figure 9.6 bears a remarkable resemblance to Ptolemy's representation of the motion of a planet (e.g., Mars) about a stationary Earth. Claudius Ptolemy in fact anticipated Fourier analysis by some sixteen centuries! Each time new measurements revealed slight inaccuracies in Ptolemy's geocentric model, pre-Keplerian mathematicians added more circles riding upon Ptolemy's previous circles, in effect including higher-order terms of a Fourier series just as today's engineers and scientists might do in studying a complicated periodic motion.

Does this mean that the early geocentric model of planetary motions might have succeeded, had the mathematical astronomers just been more diligent about including additional circles turning upon circles? Certainly, such a geocentric model could be made every bit as accurate as the heliocentric models of Copernicus and Kepler. Unfortunately, however, an acceptable level of accuracy can be gained in a geocentric model only by building in hundreds of circles (corresponding to hundreds of sinusoidal functions). This level of complexity results from the requirement that the fundamental period in a Fourier series must equal the period at which *all* details of the observable motions repeat themselves. Because the periods of the outer planets are so long (several centuries), describing the *daily* changes in their motion requires an enormous number of terms of the Fourier series. The geocentric model grows too complicated too quickly, and I am personally unaware of anyone who has attempted the challenge of using Fourier analysis to bring it into compliance with modern standards of accuracy in space science. The heliocentric model, based on elliptical orbits around the sun rather than circular orbits around Earth, is mathematically more direct.

Note, however, that this does not prove that the geocentric model is wrong, or that the heliocentric model is right. All we can say (in the spirit of Occam's razor) is that the heliocentric model is the *better* description of planetary motions, because it predicts future motions from the fewest number of independent parameters. Today, in fact, relativity theory tells us that the planets orbit neither Earth, nor the sun, nor anything else in particular. Yet, although we are free to describe physical phenomena in any reference frame we choose to use, we find that heliocentric conic-section orbits are mathematically connected to geocentric Fourier-series orbits. Surely we cannot dismiss such deep mathematical links as mere artifacts of our human subjective biases. Instead, the formal interconnections between alternative mathematical theories of the universe suggest that, while absolute truth may not reside in any specific mathematical system, physical truths are indeed intimately connected to mathematical reasoning.

Cycloids

We've seen that the simple pendulum is isochronous only for small amplitudes. This discovery troubled Christiaan Huygens (1629–1695), who among his many interests sought to build a pendulum-regulated clock whose timekeeping accuracy would not be affected by variations in amplitude. The problem is particularly troublesome for spring-driven (rather than weight-driven) clocks, for when a clock spring is wound tightly, it transmits a sizable torque to the escapement, which kicks the pendulum into a large-amplitude oscillation. Later, as the spring runs down, the impulses from the escapement are less energetic, and the pendulum's amplitude of oscillation decreases. As we've seen in Table 9.1, this change in amplitude affects the pendulum's period; if, for instance, the amplitude decreases from 15° to 10°, the period of a 1-meter pendulum decreases by 0.0049 seconds. As a result, a clock runs faster when its pendulum has a smaller swing. Each hour, such a clock will gain about 9 seconds at the lower amplitude, and in a day it will gain about 3.5 minutes. To Huygens, this degree of error was unacceptable.

Huygens discovered, however, that an isochronous simple pendulum *can* be built if the pendulum bob can be constrained to swing in a cycloidal rather than a circular arc.[8] A cycloid is the curve traced by the motion of a fixed point on the circumference of a rolling wheel (Fig. 9.7). The cycloid was apparently known to the ancient Greeks, and it was certainly understood by 1501; in the following two centuries, mathematicians worked out its area, its volume of revolution, and other properties. Huygens proved mathematically that a pendulum that swings in a cycloidal arc should be isochronous. But how does one physically achieve such a motion? An unusual and convenient property of the cycloid came to Huygens's aid: if we roll a line over a cycloid, a fixed point on that line traces out an identical cycloid. A pendulum suspended on a string whose motion is constrained by two symmetrical cycloids as in Figure 9.7 is the mechanical analog of this geometrical relationship.

There are, however, practical difficulties with such a physical arrangement. It is one thing to prove, using points and lines, that the involute of a cycloid is an equal cycloid; it is quite another problem to apply this mathematical result to a mechanical arrangement where pendulum suspensions have finite thickness. As we've seen, very small variations in a pendulum's

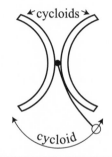

Figure 9.7. Huygens's cycloidal pendulum. The involute of a cycloid is an equal cycloid.

period can quickly accumulate to a significant error in a clock's timekeeping accuracy. Although Huygens's cycloidal-pendulum clocks did tick, the very clever mathematics he employed in their design did not prevent them from accumulating errors, and after this additional complication in their construction their lack of accuracy proved disappointing.

Ultimately, the problem of the anisochronism of the simple pendulum was solved by using power sources that provided a constant dribble of energy to the escapement, rather than by reinventing the pendulum itself. By keeping the pendulum swinging at a constant amplitude, a steady power source assured that the period of oscillation remained constant. Through this approach, pendulum clocks of impressive accuracy were available by the early 1800s.

Yet just as the cycloid was disappearing from timekeeping devices, the demands of the Industrial Revolution drove it to appear in another application: gears. Gears themselves are a very old invention; they were described by Vitruvius in the first century C.E., and were certainly used several centuries earlier. With the new machinery of the nineteenth century, however, gears needed to spin much faster than ever before. At high speeds, any sliding friction at the contact points between pairs of meshing teeth will generate heat, accelerate wear, and contribute to annoying if not catastrophic failures.[9]

Is there any way to shape a pair of gears so that when one drives the other, the teeth roll on each other without any sliding? Clearly, the teeth of such gears should not be cut into square or triangular shapes, but in curves. And because the cycloid is its own involute (Fig. 9.8), a pair of meshing cycloids can indeed be configured to roll over one another with no slipping. As a bonus, it is possible to space cycloidal gear teeth so that as one pair of teeth disengages, the next pair has already engaged.

Again, however, we confront the reality that the physical world does not permit perfect geometrical shapes, cycloids included. Gears can be machined only to finite tolerances, and even after manufacture their shapes change with temperature. As a result, gear trains tend to "chatter" due to intermittent loss of surface contact, and more so at higher speeds. This chattering effect is injurious to the mechanism, particularly near the tips of the teeth where they disengage. To minimize chattering, gears in modern high-speed applications are machined in a three-dimensional geometry, with their cycloidal sections wrapped around the gear hub in a helix, as shown in Figure 9.9. By assuring that multiple pairs of gear teeth are

Figure 9.8. Cycloidal gear teeth transmit forces by rolling rather than sliding.

always engaged simultaneously, such helical gear trains smooth out the transmission of force from one gear to the other.[10]

Euler's Equation

We've seen that π is an irrational number, not expressible as a ratio of integers. In 1882, π was also proved to be transcendental, not expressible as the solution to any algebraic equation. This discovery explained why no purely algebraic computation can ever yield π as its solution, and why the only successful efforts to compute π numerically had always been based on clever combinations of geometry, successive approximations, and infinite series.

Even within a limited range—say, between 0 and 10—there exists an infinity of transcendental numbers, which renders it quite impossible to tabulate them as we might generate a list of prime numbers. There is, however, one transcendental number other than π that plays an important role in the mathematical description of the physical world. This number, sym-

Figure 9.9. Helical gears. The teeth themselves are still cycloidal in cross section.

bolized by e, was proved to be transcendental in 1737, some thirty years before the proof that π is also transcendental.

The fundamental definition of e is

$$e = \lim_{n \to \infty} \left(1 + \frac{1}{n}\right)^n,$$

and it can also be expressed through the infinite series

$$e = 1 + \frac{1}{1}! + \frac{1}{2}! + \frac{1}{3}! + \frac{1}{4}! + \cdots \frac{1}{n}! + \cdots$$
$$= 2.718\ 281\ 828.\ldots.$$

This number e is found in the natural laws of growth and decay. A population of fruit flies, for example, grows at a rate that depends on the current population: the more flies we have, the greater the number of baby flies that will swarm around us tomorrow. If we write and solve the differential equation for this population growth, we arrive at the same conclusion as we would if we counted the fly population day by day: any initial population N_o grows to a population N according to

$$N = N_o\, e^{kt},$$

where t represents time and k is a net reproduction rate.

In a reverse example, the temperature of a hot object relative to its surroundings, ΔT, is observed to decay in time at a rate that depends on the

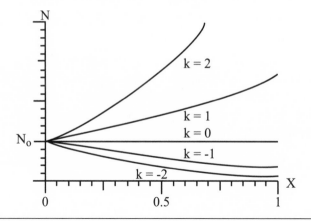

Figure 9.10. The function $N = N_o\, e^{kt}$, for positive and negative values of k.

current temperature difference. Here, the ΔT decreases from its original value ΔT_o according to

$$\Delta T = \Delta T_o\, e^{-kt},$$

where k is the decay rate. The formal similarity between these two equations suggests that growth and decay are essentially the same process, differing only in the sign of the growth-rate parameter k (Fig. 9.10).

Meanwhile, the number i (defined as $i = \sqrt{(-1)}$) requires us to leap intellectually beyond the bounds of even the transcendental numbers. Obviously, there is no measurement we can make that, when squared, will give us a negative answer. By itself, i cannot carry any physical message. Attributing physical meaning to i is not the responsibility of mathematicians, but rather a task for scientists and engineers who might want to employ the mathematics of complex variables as a physically meaningful analytical tool.

In fact, today's engineering and scientific communities routinely calculate with complex numbers of the form $c = a + ib$, using operational definitions for how the real and imaginary parts are to be measured independently. Any complex number c conforms to standard algebraic rules, so that, for instance, $c^2 = (a + ib)(a + ib) = a^2 + i^2 b^2 + 2ib = (a^2 - b^2) + i(2b)$. Here, the real part of c^2 is the quantity $a^2 - b^2$, while the imaginary part of c^2 is the quantity $2b$, and each part may be assigned a different physical meaning. In electrical engineering, for instance, the imaginary part of a circuit's impedance is called the reactance, while the real part is called the resistance. Both the reactance and the resistance can be measured independently, and both are relevant to the transmission of electrical power.

The prolific Swiss mathematician Leonhard Euler (1707–1783) discovered that the numbers e and i are intimately linked to trigonometric sines and cosines, and therefore to circles. Using x to represent any arbitrary real number, he found that

$$e^{ix} = \cos x + i \sin x$$

and therefore that $\cos x$ may be represented as $Re(e^{ix})$, while $\sin x$ is $Im\,(e^{ix})$. Accordingly, the equation of simple harmonic motion, $y = y_m \cos \omega t$, may alternatively be expressed as $y = Re(y_m e^{i\omega t})$. This complex-variable approach leads to significant computational simplifications when applying Fourier analysis or solving differential equations that describe combinations of harmonic motions.

Beyond these computational applications, Euler's equation reveals a curious relationship, beautiful in its simplicity. Setting $x = \pi$ in the equation $e^{i\pi} = \cos \pi + i \sin \pi$, we arrive at

$$e^{i\pi} + 1 = 0.$$

In other words, the five most-fundamental numbers in applied mathematics $(0, 1, i, e,$ and $\pi)$ are intimately connected!

Indeed, in the physical world, we find over and again that discrete objects (described by the number 1), growth and decay (described by e), circular geometries (π), and hidden variables (i) are holistically related as they stand in opposition to nothingness, symbolized by the mathematical zero. Although modern science still falls short of offering us a comprehensive grasp of how the universe is interleaved, pockets of mathematical description do give us insights into nature. The fact that $e^{i\pi} + 1 = 0$, each quantity directly connected to a set of fundamental physical observations, suggests a level of validity to the Pythagorean notion that the workings of nature are intrinsically mathematical.

CHAPTER 10

WAVES

On September 19, 1985, the buildings within a 10-square-mile section of Mexico City began to shudder. Within the next two minutes, 800 of these structures collapsed, killing more than 10,000 people, injuring another 50,000, and leaving around 250,000 homeless. About 58 seconds earlier, a geological fault 350 km (220 mi) to the west had suddenly slipped several meters. Clearly, the destruction in Mexico City resulted from the earthquake on Mexico's Pacific coast, even though these two events were widely separated geographically.

In Aristotle's philosophy of nature, causes needed to be proximate to their effects. If we look out a window and we see someone's hat fly off, for instance, an application of Aristotelean logic drives us to conclude that there must be a wind outside even though we can't see the moving air. Yet there are other natural phenomena, not addressed in Aristotle's writings and not well understood in those early times, that strain the notion that causes and their effects must be in close proximity. Although Aristotle did point out that events can result from chains of causes, his requirement of proximity was incapable of logically accounting for action at a distance.

Earthquakes are but one example of nature connecting widely separated events. Planet Earth gets its surface warmth from a star 93 million miles away, and if the sun were to suddenly wink out, we would not learn about this unhappy event for about 8.3 minutes (the time it takes light to travel from the sun to Earth). Meanwhile, the sun's gravity tugs on Earth across the same 93 million miles of empty space, and Earth is attracted not to where the sun *is*, but to where the sun *was* about 8.3 minutes ago. Similar examples of action at a distance abound. It is through such actions that our world, and indeed our entire universe, is bound together.

But do phenomena as diverse as earthquakes, gravity, and light have anything in common? Indeed they do. Their similarities are revealed through mathematical descriptions of them as wave phenomena.

Wave Motion

Nature rebels against highly localized concentrations of energy. A splash in the ocean or a thump on a drum dies out as the energy is carried away by oscillations induced in the surrounding medium. Such oscillations, which propagate through space and time, are called *waves*:

> A wave is a disturbance that travels through a medium in such a way that (a) at any instant in time, some physical variable is a function of the space coordinates, and (b) at any point in space, the same variable is a function of time.

There are many different physical variables that support wave motion. With a water wave, the relevant variable is the height of the water surface; with a sound wave, it is the air pressure; with a light wave, it is the strengths of a pair of coupled electric and magnetic fields. For our purposes here, we'll symbolize all such "waving" variables by w. For a wave traveling in the x-direction, w is a continuous function of the form $w = f(x \pm vt)$, where v is the wave speed—the speed of propagation of the wave pattern.

If the wave speed is constant (and it will be, if the physical parameters of the medium don't change), the waving variable w is a solution to the following partial differential equation:

$$v^2 \frac{\partial^2 w}{\partial x^2} = \frac{\partial^2 w}{\partial t^2}.$$

This equation links a wave's variations in space to its variations in time, through the constant v that represents the wave speed. An infinite number of functions satisfies this equation, and indeed we find empirically that there are an infinite number of physically possible wave shapes. The waveform need not even be periodic; it can be a pulse, for instance. It turns out, however, as we shall soon see, that even the most complicated wave shapes,

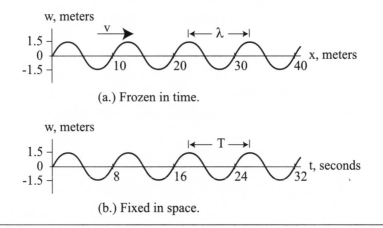

Figure 10.1. A "snapshot" of a simple sine wave traveling to the right. Here, the wavelength λ is 10 meters, the period T is 8 seconds, and the amplitude w_m is 1.5 meters.

pulses included, can be reduced to superpositions of simple circular functions: sines and/or cosines.

Let's look at the pure sine wave in Figure 10.1, which could be a water wave or a wave in a long stretched string. This wave was created at some time in the past, and it travels to the right. Diagram (a) is a "snapshot" that freezes the wave at a fixed instant in time, and diagram (b) shows how this same wave's displacement varies in time at a fixed point in space. The wave can be described by the function

$$w = w_m \sin[k(x - vt)],$$

where w_m is the maximum value of w (referred to as the wave's *amplitude*), and k is a constant. Clearly, w_m for this wave is 1.5 meters. To find k for this wave, we note that the pattern repeats its shape at regular intervals of 10 m in the x-direction, whereas the sine function repeats itself when its argument increases by 2π (radians). Therefore,

$$\sin[k(x + 10 - vt)] = \sin[2\pi + k(x - vt)],$$

which tells us that

$$k\,(x + 10 - vt) = 2\pi + k(x - vt),$$
$$10k = 2\pi,$$

and

$$k = \frac{2\pi}{10}.$$

With the constant k now determined, and given that the wave amplitude w_m is 1.5 m, the mathematical expression for this wave becomes

$$w = 1.5 \sin\left[\frac{2\pi}{10}(x - vt)\right],$$

where x is in meters, t is in seconds, and v is in m/s.

Can we derive a numerical value for the wave speed v in the same manner? Yes. Looking at Figure 10.1(b), we note that at a fixed point in space, the wave repeats itself in a time interval of 8 seconds, which again corresponds to the 2π-radian periodicity of the sine function. The algebra follows the same logic as before, leading to the result that $v = 1.25$ (m/s). This result allows us to describe our wave even more specifically:

$$w = 1.5 \sin\left[\frac{2\pi}{10}(x - 1.25t)\right].$$

Substituting any specific time t and position x in this equation gives us the wave's height or displacement at that time and position.

With this example behind us, we can streamline the process of determining the numerical constants for a particular wave. We note that 10 meters is the distance between consecutive crests of the wave in Figure 10.1. This quantity, called the *wavelength*, is symbolized by λ (lambda). We also note that if we fix ourselves at some position x_o as the wave passes, we will observe one complete oscillation in a time interval T, the wave period. The wave speed v has no choice but to assume the value of the wavelength divided by the period ($\lambda \div T$, or distance $\div$ time).

Scientists often find it convenient to describe wave phenomena in terms of constants other than v, T, and λ. Definitions of these alternative quantities and their expressions for a simple sine wave are summarized in Table 10.1. All such formulas, it should be noted, describe the very same wave phenomena, and all of these quantities are interrelated algebraically.

Of particular interest is the quantity ω, the angular velocity, usually expressed in units of radians per second. Because ω implies a circular geometry and rotational motion, it's fair to ask whether anything in a wave is

Table 10.1. Alternative but equivalent expressions for a simple sine wave propagating in the positive x-direction at a speed v.

$$w = w_m \sin[k(x - vt)]$$

$$w = w_m \sin(kx - wt)$$

$$w = w_m \sin\left[2\pi\left(\frac{x}{\lambda} - \frac{t}{T}\right)\right]$$

$$w = w_m \sin\left[2\pi\left(\frac{x}{\lambda} - ft\right)\right]$$

Variables:
  x = position;
  t = time;
  w = wave displacement at a particular position and time.
Constants:
  w_m = wave amplitude = maximum value of the wave displacement w;
  λ = wavelength = horizontal distance in which the wave repeats itself;
  T = wave period = time in which the wave repeats itself;
  v = wave speed = λ/T;
  k = wave number = $2\pi/\lambda$;
  f = wave frequency = number of waves passing a fixed point in a unit of time = λ/T;
  ω = angular velocity = $2\pi f = 2\pi/T$.

really rotating. If we watch water waves approaching a beach, our visual image is that the wave pattern moves forward while the water itself oscillates up and down, until the depth grows so shallow that the wave begins to break and the sine function no longer applies. Prior to this breakdown, the sine function tells us that the vertical displacement of the wave (as measured from its equilibrium level) oscillates just like a mass bouncing up and down on a spring.

But vertical displacement is an abstraction, while water is real. The easiest way to see what the water itself is doing, at least at the surface, is to toss a floating object onto a wave and watch its motion. When we do this, we find that the surface does not move up and down, but rather in an approximate circle (Fig. 10.2). The quantity ω turns out to be the rate of angular rotation of the individual water droplets. For water below the surface, such observations can be conducted in a laboratory apparatus called a wave tank.

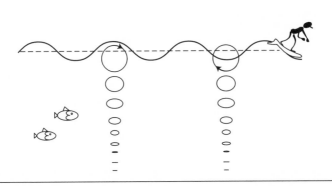

Figure 10.2. The surface water in a wave moves in a circular path at an angular velocity $\omega = 2\pi/T$, where T is the period of rotation. Deeper water moves in ellipses of decreasing size and increasing eccentricity.

Here, by viewing a wave from its side, we can establish that the deeper water moves in ellipses, which get flatter and smaller with increasing depth.

Finally, in our efforts to connect real waves with mathematical descriptions of them, we need to note that physical wave speeds cannot be altered by changing the period or wavelength. Wave speeds are determined by the properties of the medium: sound travels through air at 343 m/s, light waves streak through space at nearly 300,000 km/s, water waves travel at a speed that depends on the depth of the water, and earthquake waves propagate at speeds that depend on the properties of the bedrock. We cannot make such waves travel faster or slower by changing their ω or T. Instead, as the wave period T is increased, we find that the wavelength λ increases in direct proportion, and the wave speed, λ/T, remains constant.

Fourier Analyses of Waves

We've examined, in chapter 9, how the superpositions of simple sinusoidal oscillations can produce more complicated patterns of oscillations. We've also seen how the inverse mathematical operation of Fourier analysis can reduce any complicated oscillation into a sum of its simple sinusoidal components, each with a different period and amplitude.

With waves, we have a phenomenon that oscillates both in time and in space. At first inkling, it might seem that this would considerably complicate any mathematical attempt to describe the superposition of waves. As a practical matter, however, we can Fourier analyze complicated wave shapes either by freezing them in time or by freezing them in space. In the time domain, this gives us the wave's frequency components, while in the space domain we get the corresponding spectrum of wavelengths. These two approaches can each stand on their own, one being transformable to the other, because the product $f\lambda$ is a constant (the wave speed). An example is shown in Figure 10.3. In the top left illustration, showing a wave's temporal dependence at a fixed point in space, a Fourier analysis reveals that the wave is composed of a superposition of simple sine waves that have the frequency spectrum shown just below. In the top right-hand illustration, we see the same wave as it appears frozen in space, and below, this wave's wavelength components. Notice that the longer wavelengths correspond to the lower frequencies (i.e., the longer periods), and that the wavelength spectrum can be computed directly from the frequency spectrum by noting that for each harmonic component $\lambda = v/f = vT$. In this particular case, if λ_o is 3 meters and f_o is 20 cycles per second, the wavespeed v is 60 m/s, and this holds for all of the wave's Fourier components.

Although we can approach such Fourier analyses in terms of either frequency or wavelength, most commonly we do so in terms of frequency. This allows us to use the same mathematical description for wave spectra as we did in chapter 9 for oscillating systems (a pendulum, for instance) that do not propagate through space.

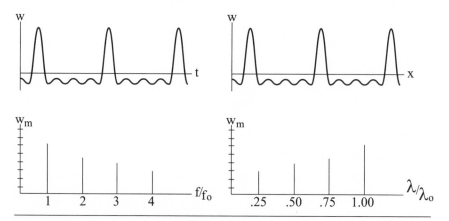

Figure 10.3. Fourier analysis of a periodic wave reveals either its frequency components or its wavelength components.

Wave phenomena in nature, however, are sometimes far from periodic. A periodic mathematical function extends in space from $-\infty$ to $+\infty$, and in time from $-\infty$ to $+\infty$. Any physical wave, in contrast, must have a beginning and an end, and therefore deviates from periodicity to at least this extent. As a result, the spectrum of a real wave never reduces to a discrete set of exact frequencies, but rather to some continuous distribution of frequencies.

The most extreme instance of a nonperiodic wave is a single pulse. To find the frequency spectrum of a pulse, we pretend that it is actually a periodic sequence of such pulses, separated by an infinite time interval. The corresponding Fourier series requires the summation of an infinite number of sinusoidal functions whose frequencies differ by only infinitesimal amounts. In the language of calculus, the sum becomes an integral, and the discrete frequencies of a Fourier series merge into a continuous frequency distribution of a Fourier integral. Figure 10.4 shows the mathematical results. Here, we begin with a pure sinusoidal wave, and find (as usual) that it is characterized by a single frequency. We then shrink the wave so it has a beginning and an end; as a result, the single frequency broadens

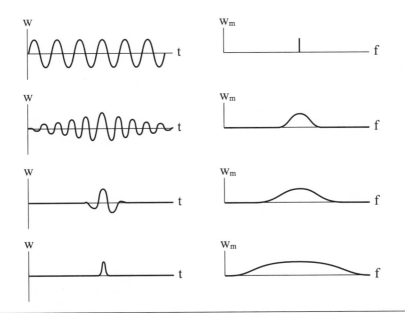

Figure 10.4. Fourier integrals. As a wave shape becomes nonperiodic, its frequency spreads into a band of frequencies.

into a small range of frequencies. We shrink the wave further, and the frequency spectrum broadens further. Finally, we shrink the wave to a narrow pulse; here, the wave's frequency components are found to span a very broad range.

The results shown in Figure 10.4 derive from mathematical logic that involves infinities and infinitesimals. Do such continuous bands of frequencies really exist *physically*? Indeed they do. The most accessible example is found with radio waves. If we listen to an AM radio during an electrical storm, we're sure to hear bursts of "static," regardless of the frequency our receiver is tuned to. Lightning discharges create powerful and narrow electromagnetic wave pulses, whose frequency spectra are therefore very broad. Regardless of the frequency our receiver is tuned to detect, that frequency is sure to be present. Other more controlled experiments verify that such predicted frequency spectra are also observable with sound waves, light, water waves, and even seismic waves. Nature indeed seems to have structured complex physical wave phenomena in a manner consistent with a human system of mathematical logic based on the geometry of the circle. To this point we shall return later.

Plane, Circular, and Spherical Waves

I limited the previous discussion to waves that travel in a single direction. Space, however, is three-dimensional, and a wave is capable of traveling in all directions through its medium. Figure 10.5 shows a wave that we might observe after tossing a pebble into a pond. This wave radiates outward from its source in a pattern of concentric circles whose amplitude decreases as the wave front circumference increases. At great distances from the source, the angular dispersion of this circular wave becomes less noticeable, and small cross sections of the wave appear to propagate linearly.

Similarly, a wave in three-dimensional space spreads out in a spherical pattern. Here, the wave amplitude decreases as the *surface area* of the wave front expands with increasing distance from the source. This is impossible to show on a two-dimensional diagram, but it follows by analogy to Figure 10.5. Sunlight striking Earth, for instance, has reached us by spreading out spherically from its source. By measuring the intensity of the sunlight at Earth's surface, it becomes an easy matter to work backward through the spherical wave geometry to calculate the intensity of the radiation at

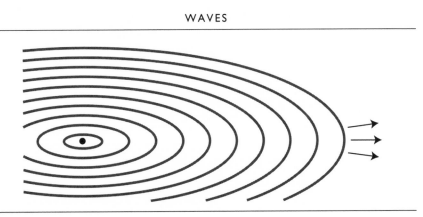

Figure 10.5. A circular wave generated by a point source. At large distances from the source, the wave's angular dispersion is less apparent.

the surface of the sun. In this manner, we gain scientific information about solar processes that are taking place 93 million miles away.

The geometrical spreading of waves also has observable consequences closer to home. Underwater earthquakes and explosions of sea-level volcanoes often generate long-period sea waves called *tsunamis*. Near their source, tsunamis can have tremendous amplitudes, 30 meters (100 ft) or more; they have been known to totally obliterate coastal settlements, and even to beach steamships several kilometers inland.[1] As the distance from the source increases, however, the amplitude of a tsunami decreases at a rate predicted by its expanding circumference. Similarly, the expanding spherical waves from atmospheric explosions decrease in amplitude consistent with the increase in surface area of the expanding wave front.

One of many practical examples is locating the epicenter of an earthquake. An earthquake generates two types of waves: a P-wave (primary wave), which is essentially a large-amplitude sound wave that propagates through shallow bedrock at a speed of about 5,400 m/s, and an S-wave (secondary wave), a transverse vibration similar to a water wave that propagates through bedrock at about 3,200 m/s. Both types of waves expand from the earthquake source as circular wave fronts. If we are close to the source of an earthquake, we detect these two waves nearly simultaneously. If we are farther away, however, we first notice a rumbling vibration from the floor, followed a few seconds later by a destructive lateral shaking as the S-wave strikes us. The difference in time of arrival of the P- and S-waves is about 12.7 seconds for each 100 km from the source (the reader can confirm this figure from the above wave speeds). On the basis of a measured

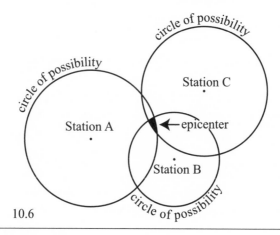

Figure 10.6. Data from three seismological observatories can be combined to locate the epicenter of an earthquake.

Δt, then, a single seismographic observatory can easily determine its distance from the source of an earthquake.

Yet knowing the distance from an observer does not pinpoint an earthquake's epicenter. If an epicenter is, say, 200 km from us, all we know is that the earthquake occurred somewhere on a 200-km circle of possibility centered at our point of observation. To pinpoint the epicenter itself therefore requires a minimum of three separate observations, as shown in Figure 10.6. Note that the circles shown in this diagram are not the actual wave fronts, but are instead loci of possibilities. Only when three of these circles of possibility intersect does possibility become reality. It is this intersection point that both physically and mathematically identifies the source of the expanding P- and S-waves of an earthquake.

Space, Time, and Light

Light is a wave phenomenon, and a most ubiquitous one. A large percentage of our observations of the universe depend on this phenomenon; we learn of reality largely through what we see. Yet in a sense we are all mostly blind, for the human eye responds only to wavelengths within the range of about 400 to 700 nanometers. Shorter wavelengths (those beyond the violet end of the visible spectrum) include ultraviolet light, X-rays, and gamma rays, while wavelengths that are too long to be visible (those beyond the red end of the spectrum) include infrared radiation, microwaves, and radio waves. This entire spectrum of wavelengths, from gamma rays at the short end to radio waves at the long end, is called the *electromagnetic spectrum*. Visible light constitutes but a tiny fraction of all possible electromagnetic waves.

We group all these electromagnetic waves together because they have much in common. Three properties are of particular interest: (1) Electromagnetic waves do not require a physical medium to carry them. They travel quite well, even over cosmic distances, through the vacuum of empty space. (2) In a vacuum, all electromagnetic waves travel at the same speed, 299,792,458 m/s. (3) All electromagnetic waves are produced by moving (or oscillating) electrical charges or magnets. These properties were well known by the 1890s, due to the mathematical theory of electromagnetism developed by James Clerk Maxwell, the experimental discovery of radio waves by Heinrich Hertz, and the failed herculean efforts of A. A. Michelson and E. W. Morley to detect the "ether," a postulated medium that might carry light through what only appears to be empty space.

Albert Einstein (1879–1955) was not a man who relished the messy world of experimental physics, where measurements are always a bit fuzzy and equipment can be cantankerous. He much preferred the *Gedanken*, or thought, experiment. One of his early thought experiments was to imagine himself traveling on a railroad train at the speed of light. He would look back and view the crowd in the station he'd just whizzed through. Would he see anything unusual? Einstein thought so. But to figure out just what, he could no longer depend on verbal logic alone, and he certainly couldn't conduct a physical experiment.

Prior to this time, Einstein was not a particularly talented mathematician; one of his professors, Ernst Mach, had actually told him that he would never amount to anything. In 1905, Einstein found himself in a ho-hum job as a patent clerk, but with an intellectual curiosity that drove him to try to conduct his Gedanken experiments through mathematical logic. In the tradition of Euclid, Einstein structured his analysis of high-speed motion on two postulates. These, however, were not mathematical but rather physical postulates, intended to refer to the real and observable universe. Taking some slight liberties with Einstein's own language, we may state these two postulates as follows:

(1) The speed of light is a constant, independent of any relative motion of the source or observer.

(2) If two observers are moving toward or away from each other at a constant speed, there is no experiment either one can perform to determine who is really moving, or how fast, in any absolute sense.

Are these postulates reasonable? Apparently so. The first can be tested by collecting light from a distant star that is receding from us at a high speed, directing this light into the laboratory through a telescope, then measuring the speed of the light wave. Contrary to intuition, the light is not found to travel slower because the source (the star) is speeding away from us. Light is always found to travel through space at a fixed speed, regardless of the relative motion of its source and observer, and that's that.

As for the second postulate and the impossibility of determining who is really moving, this is a more familiar notion. Suppose that we fly from Atlanta to Los Angeles (due west) in a supersonic plane that cruises at 875 mi/h. Below, we see the ground passing under us, and from our vantage point we may just as well be at rest while Earth's surface speeds to the east at 875 mi/h. So which is it: are we moving, or is the ground moving, and how fast? Clearly, it depends on our perspective.[2] To an observer on the ground, we are moving. To an observer on the moon, we appear stationary while the ground travels beneath us to the east. To an observer in an orbiting space station, both we and Earth's surface are moving in the same direction, but at speeds that differ by 875 mi/h. All are equally valid descriptions of the reality, and no physical experiment can distinguish among them.

Although we can't actually prove the two postulates, we have established their physical plausibility and the lack of physical evidence to the contrary. (In pure mathematics, this is not a requirement; in science, it is.) With these postulates in place, we can now conduct a Gedanken experiment, essentially the same as Einstein's, to examine their consequences.

Figure 10.7 shows two reference frames, xy and $x'y'$, moving relative to each other in the xx' direction at a speed v. A light bulb is at the origin of one of these frames, and at the instant the two origins coincide, this light bulb is flashed. According to the first postulate, a spherical light wave expands outward at a speed of 299,792,458 m/s from *each* origin, even though they are moving with respect to one another. Using c to represent the speed of light, this spherically expanding wave front satisfies the following two equations of analytic geometry:

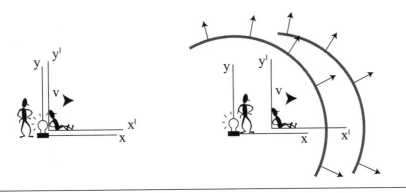

Figure 10.7. A spherically expanding light wave as observed from two reference frames in relative motion.

in the xy frame, $\quad x^2 + y^2 + z^2 = R^2$, where $R = ct$;

in the $x'y'$ frame, $\quad x'^2 + y'^2 + z'^2 = R'^2$, where $R' = ct'$.

We now seek a way to reconcile these two equations. First, it's clear that $y = y'$ and that $z = z'$, because there is no motion in these directions. This allows us to subtract the second equation from the first, to get

$$x^2 - x'^2 = c^2 t^2 - c'^2 t'^2,$$

or equivalently

$$x^2 - c^2 t^2 = x'^2 - c^2 t'^2.$$

Remember, x is the position of the expanding wave front relative to the unprimed reference frame, and x' is the position of the *same* expanding wave front relative to the primed reference frame. Obviously, $x' \neq x'$ at any time $t > 0$. It follows, then (without any calculation, but just by reading the equation), that $t \neq t'$ at any time $t > 0$.

This is truly an astounding conclusion! It says that time is not absolute, but instead that the passage of time is different for observers if they are moving relative to one another. It also follows that a position in space as measured by two different observers in relative motion is somehow determined through c, the speed of light. I won't present the details of the algebraic logic here, which are somewhat lengthy.[3] Suffice it to say that the analysis leads to the following relationships:

$$x' = \frac{x - vt}{\sqrt{1 - \frac{v^2}{c^2}}}$$

and

$$t' = \frac{t - \frac{vx}{c^2}}{\sqrt{1 - \frac{v^2}{c^2}}}.$$

These relationships are usually referred to as the *Lorenz transformation*. The interested reader can verify algebraically that they indeed satisfy the conditions of the prior equation and the original Gedanken experiment. Note that when $v \ll c$, these equations reduce to $x' = x - vt$ and $t' = t$, which is just what our everyday experience tells us they should be. Only at speeds approaching the speed of light do the time and space variables become intertwined with each other.

Let's now place a clock and a measuring rod in the x' frame, and observe them from our x frame as they whiz away from us at some speed v. We have no choice but to make these observations through light (watching through a telescope, for instance) or through some other form of electromagnetic radiation. The rod has some extension in space, and its length L_o in the x' frame (its own rest frame) is the difference $x_2' - x_1'$ of the positions of its ends. According to the Lorenz transformation, however, what we observe in watching from our frame is some *shorter* length L, where

$$L = L_o \sqrt{1 - \frac{v^2}{c^2}}.$$

Meanwhile, the clock is ticking away at some fixed point in the moving x' frame. The time interval between ticks is, say, Δt_o. What we observe from our x frame, however, is some different time interval on our own clocks, $\Delta t = t_2' - t_1'$. Here, the Lorenz transformation tells us that

$$\Delta t = \frac{\Delta t_o}{\sqrt{1 - \frac{v^2}{c^2}}}.$$

This says that as we watch the clock in the moving frame tick off one hour,

Table 10.2. Predictions of relativity. Time intervals, lengths, and optical wavelengths of a moving object, as observed from a separate reference frame, for relative recessional speeds approaching the speed of light.

v/c	Δt	L	wavelength	"color"
0	1.0000 h	1.0000 m	450.0 nm	blue
0.2500	1.0328	0.9682	580.9	yellow
0.5000	1.1547	0.8660	779.4	"near" infrared
0.7500	1.5119	0.6614	1.191 μM	infrared
0.9000	2.2942	0.4358	1.961	infrared
0.9700	4.1134	0.2431	3.647	infrared
0.9900	7.0888	0.1411	6.350	infrared
0.9950	10.0125	0.0999	8.991	infrared
0.9990	22.3663	0.0447	20.12	infrared
0.9999	70.7124	0.0141	63.45	infrared
0.99999	223.607	0.00447	201.2	microwave
0.9999999	2236.07	0.000447	2.012 mm	microwave
1	∞	0	∞	—
>1	imaginary	imaginary	<0	—

our own clocks will tick off *more* than one hour. Our perception is that everything in the moving frame is happening in slow motion.

Table 10.2 summarizes these effects of length contraction and time dilation for a range of speeds. Let's think about a rocket speeding past Earth in a straight line, at some speed *v*. On this rocket are a rod 1.000 meter in length, and a clock that ticks off one hour in time, as measured by passengers on the rocket. As far as these passengers are concerned, everything in their own reference frame is quite normal. As observed from Earth, however, things on the rocket start to appear strange at high speeds. The rod shrinks in the direction of its motion, and the clock runs slower. At a speed of 97% of the speed of light, a one-hour class lecture by a professor on this rocket would require over four hours of viewing time on Earth, even though only an hour's worth of information is transmitted. Meanwhile, the professor giving the lecture appears strange. His nose flattens into his face, and if he sits, his knees flatten into his hips. In general, as he twists toward and away from the blackboard (in slow motion), his shape seems to undulate, growing in directions perpendicular to *v* and flattening in the direction coinciding with *v*.

With these results in mind, can we answer the question of what Einstein would have observed from his seat on the Gedanken train moving at

$v = c$? Yes, provided that his speed remains infinitesimally lower than c. As far as Einstein the observer is concerned, *he is at rest* and it is the station that is speeding away from him at essentially the speed of light. The clock on the station wall is therefore frozen in time, forever after indicating that instant in time when the two reference frames coincided. The crowd in the station is also frozen in time, and everything and everyone is flattened in the manner of a Grandma Moses painting. Meanwhile, all the colors are shifted toward the red, so far toward the red, in fact, that the wavelengths are no longer visible to the human eye. The only way Einstein can "see" the scene is by collecting the electromagnetic waves with a radio antenna, then reconstructing the picture.

And at still higher speeds? At $v = c$, the light from the station is no longer detectable by any physical means, because it now has an infinite wavelength. The whole scene at the railroad station winks out of existence, and there is no longer any physical way to learn what is happening there. The station has passed over our "event horizon." We will return to this concept in the next chapter.

Relativity and Reality

Are these strange relativistic effects part of reality, or are they just an artifact of our extending the geometrical analysis of wave propagation beyond our realm of direct experience? Scientists argued this issue for years in the early twentieth century. Gradually, however, investigations into previously unexamined pockets of physical phenomena began to shed light on the issue. Amazingly, relativistic effects like contraction of length and dilation of time *do* seem to occur in nature.

Muons, for instance, are elusive and unstable subatomic particles whose rest mass is 207 times the electron's rest mass.[4] Muons can be created in high-energy particle accelerators, and if they are moving slowly, they wink in and out of existence with a half-life of about 1.5 microseconds before decaying into other particles. Even if a muon traveled at the speed of light, the distance it would move in its lifetime is only around

$$x = ct = (3 \times 10^8 \text{ m/s})(1.5 \times 10^{-6} \text{ s}) = 450 \text{ m}.$$

It turns out that muons are also created naturally, through the bombardment of the outer atmosphere by cosmic rays from outer space. When we measure this natural muon flux both at the top of the stratosphere and at

Earth's surface 60,000 meters below, we indeed find that most of the muons created at high altitudes disintegrate before they reach the ground. Numerically, however, the survival rate is greater than one would expect on the basis of an average 450-meter travel distance. In fact, at Earth's surface we detect around eight times as many muons as ought to survive in the absence of relativistic effects. The only explanation seems to be that muons actually live longer when they are traveling near the speed of light, consistent with the prediction of relativity theory. This extended lifetime gives the muons a better chance of surviving a long trip through the atmosphere and striking Earth.

From the reference frame of the muon, however, its lifetime has not changed at all; it still lives just 1.5 microseconds. But because the muon "sees" Earth rushing toward it at a velocity near the speed of light, the mathematics of relativity tells us that in the muon's reference frame the thickness of the atmosphere has shrunk to less than a thousand meters. Muons striking the ground really haven't traveled very far after all.

From the perspective of the Earth-based observer, the muon's lifetime has increased, while from the perspective of the muon, Earth's atmosphere has contracted. Which is the reality? The second postulate of relativity says that we have no physical way of deciding. Although we analyze the process differently from the two reference frames, the only observable event is the detection of cosmic-ray muons at Earth's surface. This physical observation is totally consistent with the mathematical analysis from either reference frame.

In the fifty years since the first muon observations, many experiments have been performed on other subatomic particles moving at high speeds (particularly electrons and protons). The physical findings are always consistent with the mathematical predictions of the theory of relativity. Yet there remains the question of whether protons, electrons, and muons have anything to say about humans. We might argue, philosophically, that humans are physical compositions of submicroscopic particles, and that therefore any natural laws that apply to subatomic particles must also apply to humans. But is this really the case? Again, mathematics alone cannot answer the question; we must combine mathematical prediction with direct observation.

Accordingly, several experiments have been conducted in which scientists have carried precision atomic clocks in their baggage as they took a series of commercial flights around the world, some traveling to the east (in the direction of Earth's rotation), and some to the west. Relative to the laboratory they'd left behind on our spinning Earth, the eastward and westward flights had different velocities. And sure enough, the clocks previously synchronized on the ground always reveal that east-flying travelers age a bit more than west-flying travelers, by a few microseconds. Relativity theory, although structured as an edifice of mathematical logic, has indeed succeeded in predicting subtle natural effects that apply on human scales of space, time, and motion.

Matter Waves and the Uncertainty Principle

According to modern quantum theory, waves and matter are not distinct, but instead are different aspects of the same physical reality. For instance, although light is usually described mathematically as a wave, in the phenomenon of the photoelectric effect we find that a beam of light instead behaves like a stream of particles (photons). This curious particle-like behavior of light waves was first analyzed by Einstein in 1905.

Reasoning by analogy is an important aspect of scientific inquiry. While mathematics progresses by *demonstrative* reasoning, scientific inquiry progresses through *plausible* reasoning.[5] To a scientist, mathematics does not define truth; all it does is point out plausible roads to truth about the universe. The final arbiter of truth in science is none other than nature itself.

It was in 1929 that a young doctoral student, Louis de Broglie, suggested that if light waves can behave like particles, then, by analogy, particles might sometimes behave like waves.[6] De Broglie accordingly hypothesized that any pocket of energy in space might be alternatively described as a wave or as a particle. As a particle, a physical entity is most appropriately described in terms of its energy E and its momentum $p = mv$, while as a wave, the same physical entity has a frequency f and a wavelength λ. Drawing upon Einstein's earlier analysis of the photoelectric effect, de Broglie suggested that these quantities are always related through a constant, h, such that

$$p = \frac{h}{\lambda} \quad \text{(momentum)}$$

and

$$E = hf \quad \text{(energy)},$$

where h (Planck's constant) has the value 6.626075×10^{-34} J·s, $\pm\, 40 \times 10^{-40}$ J·s.

Subsequent experiments confirmed that these equations apply equally well to electromagnetic waves, electrons and protons, and in fact every other piece of observable reality. Waves are not just disturbances in a medium, but swarms of tiny particles that carry momentum through the medium. Meanwhile, a particle like an electron is not merely a small hard object with a well-defined surface, but rather a localized wave packet that oscillates as it moves through time and space. In fact, it is the wave nature of the electron that explains why atoms don't collapse, as they surely would if the electrons orbiting the nuclei were but particles.

Not only has de Broglie's hypothesis stood the test of time, but it has also found practical application in devices such as electron microscopes. Given enough energy, the frequency of an electron wave is high enough and its wavelength short enough that it can be used to create images of objects whose dimensions are smaller than the wavelength of visible light. Today, we can probe matter with matter, magnifying objects the size of mere molecules into visible images.

Yet nature seems to place limits on how far we can see, how deep we can probe, and how much we can magnify. For very small objects, those on atomic and subatomic scales, such limits seem to be absolute. If we try to detect the position of a small particle, the only way we can do so is to bounce a wave off it. But by this action we necessarily cause the particle to recoil, so its position after the measurement is no longer the position it had before the measurement. Similarly, if we try to measure a small object's energy or momentum, or the time it occupies some observable energy state, the very act of measurement again alters the quantity we are trying to measure. In 1927, the German physicist Werner Heisenberg proposed that nature imposes a fundamental limit on the precision with which we can make physical observations. Using the symbol Δ to represent the uncertainty in the quantity that follows, Heisenberg's principle may be stated as follows:

$$\Delta x \Delta p_x \geq \frac{h}{2\pi}$$

and

$$\Delta t \Delta E \geq \frac{h}{2\pi}.$$

Here, momentum and position are conjugate variables: measuring one very precisely forces us to give up any hope of measuring the other to high precision. In other words, if we determine where a particle is in space, we cannot simultaneously tell how fast it is traveling. Similarly, energy and time are conjugate variables: if we know a particle's energy to great precision, we cannot know with any confidence how long it retains this energy.

Initially, Heisenberg proposed his uncertainty principle as a statement of the minimum possible level of interaction between the observer and the object observed. Because h is a very small number, the uncertainty principle places no practical restriction on our everyday observations and measurements on human scales. This principle does, however, remind us that no observation can be made without affecting, to at least some small extent, the very physical variable we are trying to measure or its conjugate variable.

Before long, scientists began to examine the uncertainty principle from another perspective. Is there any difference between saying that an object's position *cannot be measured* to a precision better than Δx, and saying that the object *does not have* a position except within the range Δx? Although some philosophers would claim there is a big difference between these two propositions, to physicists there can be no essential distinction. Reality is only what we can observe, and if we can't observe more precisely than some Δx, then we have no basis for concluding that a particle's true physical position is "really" finer than this Δx. What we cannot measure, we cannot possibly confirm as reality.

If we combine this idea with the concept of de Broglie's wave-particle duality, we arrive at a strange but fascinating picture of matter at its most fundamental levels. In Figure 10.8, we represent a subatomic particle (an electron, for instance) traveling along as a small wave packet rather than as the more common billard-ball image. In the first case this wave packet is very tight, so the position of the electron is fairly well defined at each instant in time. Meanwhile, however, the momentum of the particle spans a spectrum of possibilities. When this electron interacts with another particle, the observed outcome is indeed consistent with its momentum spanning a range of values rather than having a single value. In the other case, we have a similar particle whose momentum is more well defined; now we find, however, that the particle's position grows more fuzzy in space. In some cases, a particle's "fuzziness" can even penetrate barriers, allowing the particle to "tunnel" out of its container, a phenomenon that finds application in tunnel diodes.

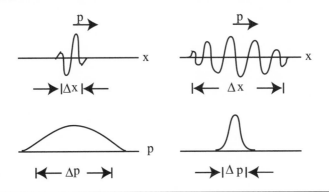

Figure 10.8. Matter waves. Particles with well-defined positions have indefinite momentum, and particles with well-defined momentum have indefinite positions.

The similarities between Figures 10.4 and 10.8 do not seem to be an accident. Just as an ordinary water wave has a broad frequency spectrum when the wave is more localized in space, a small particle will have a broad momentum spectrum when it is concentrated in a small Δx. This analogy extends to the mathematics. The Fourier integral can be modified slightly into an operator known as a Fourier transform, which allows us to calculate the momentum distribution function from the position distribution function, and vice versa.

We note, through all of the mathematics of waves, matter waves, and the rest of quantum theory and atomic physics, that the quantity π keeps reappearing. Let's briefly trace how this happens, at least in terms of the sequence of the mathematical logic. Pi was first associated with the circle, then with angles (2π radians in one revolution). As a result, functions that are defined in terms of circles (e.g., the trigonometric sine and cosine) necessarily repeat themselves when their argument increases by 2π. Because complicated wave shapes can always be reduced to sums or distributions of sine and cosine functions, the analysis of such wave shapes also involves π. And finally, because matter itself has wave properties, π also enters the mathematical description of the fundamental particles and their interactions.

This sequence of logic notwithstanding, it still seems very curious that π should be so intrinsically intertwined with the properties of matter, when initially it arose as an abstraction that described ideal circles. We will further examine this deep connection between mathematical logic and the structure of the physical universe in the last two chapters.

CHAPTER 11

ARTIFICIAL AND NATURAL STRUCTURES

Even in earliest times, humans often sheltered themselves with vertical walls supporting horizontal beams that, in turn carried a roof. Simple horizontal beams are also an age-old means of building short bridges. Because, however, horizontal beams always sag somewhat, even under their own weight, their span is always limited. Predicting how a beam will behave in an application requires us to understand the interplay between the geometry of a structural element and its natural material properties.

In Figure 11.1, we see that tension forces arise in the convex side of a bent beam, while compression forces arise in the concave side. Beams fabricated from wood or steel can tolerate a fair amount of bending, because they are equally strong whether in tension or in compression. Masonry materials like stone, brick, or unreinforced concrete are a different story. Such materials are strong only in compression, and they easily fracture when placed in tension. As a result, the strength of a horizontal stone beam is limited by its low-strength tension side, and its convex face will fracture well before its stronger compression side is close to failing.

The ancient Egyptians and Greeks were well aware of this shortcoming of building with stone. Yet stone also offers advantages, most notably its relative permanence. If a stone structure can be designed so that no stone ever bends appreciably, that structure has a chance of surviving several millennia. In fact, the Greek Parthenon, completed in 432 B.C.E., survived intact until 1687, when the center portion was destroyed by an explosion during a Venetian attack on Turkish-occupied Athens.

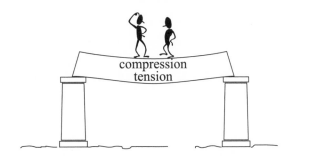

Figure 11.1. A horizontal beam sags under load. Tension forces arise in the convex side, while compression forces arise in the concave side.

The ancient Greeks compensated for the tensile weakness of stone by constructing their public buildings with many closely spaced vertical columns supporting stubby stone slabs over relatively short spans (Fig. 11.2). The roof was carried upon this network of horizontal beams, which were subject to relatively small bending stresses because of the close spacing of their supports. The stone columns themselves were pure compression members, and therefore quite capable of supporting large vertical loads. In tramping today through the ruins of ancient Greek public buildings, one can't fail to notice that the interior space was also filled with these vertical supporting columns; one did not find much open space within a Greek temple. To bring large groups of people together, the Greeks went outside, to amphitheaters.

It seems curious that the ancient Greeks, with their knowledge of geometry, did not progress beyond column-and-beam construction. Maybe they

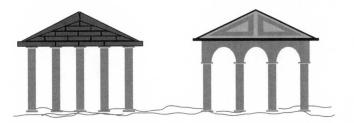

Figure 11.2. Ancient Greek and Roman architecture was based on the vertical column. The Greeks bridged the columns with horizontal slabs, while the Romans used arches.

grew so familiar with their traditional architecture that they found any deviation to be aesthetically jarring. Or maybe it was a matter of the sociology of power: those few with the authority to approve deviations from established standards were more concerned with politics than with engineering. Many Greek public buildings tumbled to the ground during earthquakes in that seismically active region of the world, even as new structures were being built according to the same basic design.

Although history does not record who discovered the principle of the circular arch, we do know that it was the Romans who first applied the arch on the scale of large structures. With the arch, no stone is ever bent, and (barring major earthquakes) every part of each stone remains forever in compression. Romans found the arch to be so successful that they built grand aqueducts and bridges as strings of arches upon arches upon arches. Many of these structures (Fig. 11.3), dating from as early as the first century C.E., remain intact today. The circular Roman arch does, however, present a design challenge, insofar as it does not transfer the load verti-

Figure 11.3. A Roman aqueduct near Segovia, Spain.

cally to its supporting columns, but instead generates a component of outward thrust that acts to tip the columns laterally. In a series of arches, this effect cancels out in all but the end arches. At its ends, then, an arched bridge must be buttressed by immovable bedrock. In a large building, the outward thrust of the arches can be minimized (although not eliminated) by building the floor plan in a circular or elliptical shape, so there isn't any end arch.

The Romans carried the arch principle this far, but no farther. The techniques needed to construct arched structures became bureaucratized, and Roman builders locked themselves into thinking about semicircles, and semicircles alone. One of the practical problems of arch construction is that the arch is not self-supporting until the last stone (the keystone) is in place. If this stone does not fit properly, the entire edifice collapses. Prior to the fitting of the keystone, an arch must be supported by temporary timber falsework. The practical limits on how big one can build such falsework set limits on how great a span one can achieve with a stone arch. Given a broad valley to bridge, the Romans were forced to resort to a large number of small arches rather than a small number of large arches. This required a great investment in labor and materials; in fact, today's engineers would characterize the Roman masonry arch as extremely inefficient in its use of materials.

By the twelfth century, builders in Europe seeking to erect ever-higher cathedrals with increased uncluttered interior space finally broke with the traditions of the Romans. A distinguishing feature of these new Gothic cathedrals was the large number of huge stained-glass windows piercing their walls. Because windows themselves are incapable of supporting any weight, it became necessary to direct the weight of the upper courses of masonry out around the windows as well as around the building's open interiors. The Roman arch failed to meet this objective, because it was much too bulky. What other geometry could be used that would still assure that no stone would ever be subjected to a force of tension?

The answer was to distort the circular arch into a shape higher than it was wide. The result was the Gothic arch, whose half-sections correspond closely to sectors of ellipses (Fig. 11.4). By 1210 C.E., using only stone Gothic arches, builders achieved open floor spaces as large as 60 ft by 180 ft with free heights up to 138 ft, the entire structure encompassing more than triple this space in its airy side vaults.[1] Walking into a Gothic cathedral today, our immediate impression is not how much stone was used,

Figure 11.4. Gothic arches are higher than they are wide, and are composed of sections of ellipses.

but rather how *little* was used, relative to the size of the open space and the huge windows in the walls.

Rigidity

We sometimes feel vibrations in a floor as someone walks past, and we may hear wood-framed buildings creak in high winds. Bridges are particularly susceptible to such movements, and most of us have experienced the sensation of a bridge swaying or vibrating beneath us when a tractor-trailer speeds past in the opposite direction. None of this is any cause for alarm, because structures *need* to deform slightly to accommodate variations in their loads. The important requirement is that they never deform so much that they cannot restore themselves to their original geometrical configurations.

Rigidity, therefore, is a relative concept, and no structural element can be perfectly rigid. Further, because physical materials can be distorted in many ways, their rigidity will depend on how they are deformed. A masonry arch, for instance, is quite rigid when loaded in a compression mode, but it fails immediately if any part of it goes into tension. For our purposes, we can consider a structure to be *rigid* to the extent that its overall geometry cannot be distorted without breaking its individual members.

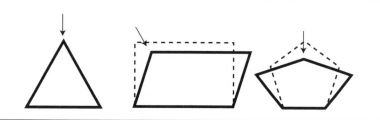

Figure 11.5. The triangle is the only rigid polygon.

In two dimensions, the triangle is only shape that unequivocally meets this requirement. There is no way to distort a triangle without stretching or compressing its sides or without breaking a corner apart. All other two-dimensional figures, from quadrilaterals to higher-order polygons, can be distorted *without* altering the length of any side (Fig. 11.5). The rigidity of physical triangles follows directly from the Euclidean theorem that a tri-angle is determined uniquely if we specify its three sides (or two sides and an included angle, or two angles and a side).

This geometrical principle long ago found its way into numerous prac-tical applications. If we want to keep a roof from sagging, we can brace it with triangles. We can design truss bridges, built of triangles, to span streams that are wider than a simple beam can reach. We survey inacces-sible terrain on Earth, and we locate distant objects in the solar system, by using the principle that triangles of light beams cannot be distorted. By measuring the accessible portions of a triangle, we thus gain knowledge of its inaccessible portions.

In three dimensions, however, additional possibilities arise. As we saw in chapter 7, Euclid proved that there are five (and only five) solids whose faces are identical regular polygons: the tetrahedron, hexahedron, octo-hedron, dodecahedron, and icosohedron (Fig. 11.6). All of these so-called Platonic solids are rigid. Many other rigid shapes can be formed, however, by slicing apexes off the regular polyhedrons. The resulting geometries are widely used in various space frames and geodesic domes.

A true dome is a hollow hemispherical structure, which may be viewed as a polyhedral surface with an infinite number of faces. The Pantheon in Rome, continuously occupied for more than eighteen hundred years, stands as impressive testimony to the durability of the dome as a structural element. Domes have since been incorporated in a large number of cathe-drals, government buildings, and sports stadiums. The dome, unlike its

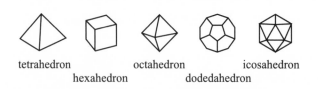

tetrahedron octahedron icosahedron
hexahedron dodedahedron

Figure 11.6. The five Platonic solids are polyhedrons whose faces are identical regular polygons.

two-dimensional cousin the arch, is inherently rigid—it cannot be deformed without altering the distances between some pairs of points on its surface. Although domes do tend to bulge outward somewhat at the bottom, this effect is not a consequence of the geometry but instead is a function of the materials used. Domes with circumferential tension members near their bases are quite inflexible. Even when a dome is built of masonry, it takes relatively little effort to counteract the outward thrust, and early builders who understood this were able to erect very successful structures.

Just as the arch can be made more efficient by distorting it into a more elliptical shape, so the dome is more efficient when it is shaped more like an ellipsoid (an ellipse rotated around an axis). In Gothic architecture, we often see sectors of such ellipsoidal domes in the vaulted ceilings. Because these regions can be viewed as triangles drawn on spheres or ellipsoids, they are referred to as spherical or elliptical triangles. It does not take a complete dome or even a major portion of it to ensure rigidity; even relatively small spherical and elliptical triangles can significantly enhance a structure's rigidity.

Amazingly, the grand Romanesque domes, the soaring Gothic cathedrals, and the earliest iron bridges were all built through an empirical approach to design. The relevance of geometry was, of course, assumed, and clearly the properties of the materials themselves could not be ignored. Yet the confidence to pursue all of these early projects grew out of a trial-and-error process rather than a quantitatively analytical approach. A small nave would be vaulted by an arch higher than wide, and it would be successful; accordingly, another nave would be built still higher. Supporting columns would be reduced in diameter, but when they began to reveal bending fractures, buttresses would be added to the structure's exterior walls. Accordingly, the next project would incorporate buttresses from the beginning. Although the existence of underlying principles was always assumed, the idea of formulating such principles mathematically was slow in developing. In the early years, the proof was in the proverbial pudding: if a structure was successful, it meant that God or nature liked it (depending on one's religious perspective) and would probably smile with approval at similar designs in the future.

With the plethora of analytical tools available to modern engineers, we are no longer surprised to find geometry applied to structures in new and creative ways. Today we build on scales beyond the imagination of those in earlier times: the Akashi Kaikyo bridge in Japan, for instance, has a free span of 1,990 meters, and the Petronas Twin Towers in Malaysia soar 451.9 meters above their bases, even though both of these projects are situated in seismically active regions of the world.[2] The incredible investments of capital and labor in building such structures reflect our confidence that the principles of geometry and materials science can successfully be applied in ways never before attempted. Mathematics indeed extends our empirical observations into whole new realms of what is physically achievable.

Geometry and the Biological World

If asked to describe a living thing, say a flower or an ant, most nonspecialists begin with a description of its size and general shape before moving on to other features like color, smell, and behavioral characteristics. In doing so, we recognize that the geometry of an organism's surface is its most distinguishing feature. Of course, biologists don't refer to such descriptions as geometry, but rather as morphology. And in some sense there is a difference; an elephant is recognized as an elephant whether sitting or standing, and whether its trunk is raised in the air or dragging along the ground. The pachyderm's spatial geometry changes, yet its morphology remains the same.

Still, the argument can be made that biological morphology is geometry of a sort, a type of geometry that is much too complicated to simplify to idealized collections of points, curves, lines, and/or algebraic equations. Because the amount of information needed to fully describe the appearance of an elephant is no less complicated than the elephant itself, a formal mathematical description would have nothing to offer us. This does not mean, however, that the animal itself is not contained within a closed surface in space; it most certainly is. Furthermore, it is empirically obvious that an elephant's shape, or that of any other living organism, can be altered only within certain geometrical (i.e., morphological) limits.

An experienced hunter can recognize a deer in but a fraction of a second, whether the deer is facing toward him, away, or laterally. This is an amazing achievement, considering the amount of information it takes to program a computer to recognize a deer and to distinguish it from, say, a cow. Although as students we struggle with formal descriptions of simpler geometries such as ellipsoids or icosohedrons, our human brains are programmed to recognize quite complex natural geometries, from a variety of perspectives, almost instantaneously. How might we explain this impressive ability?

The human brain necessarily evolved in a manner that enhanced our species' chances of survival through many millennia of prehistory. Compared to other animals, humans are not particularly fast runners, nor particularly strong, nor very efficient swimmers, nor adept at flight, nor does our meager body hair insulate against cold. Our young take a long time to reach maturity, and therefore have a long period of vulnerability to predators before they get a chance to pass on their genes. If the human species had lacked the ability to compensate for such shortcomings, we would not be here today to ponder the question of how we recognize complex natural geometries so readily.

Two of the prime requisites for survival in a natural environment are (1) to avoid being eaten, and (2) to find things to eat. Both of these skills rest heavily upon our ability to quickly identify other organisms, prey or predator, through their morphology. To respond to the threat of a wolf in ages past, we needed to recognize this animal, whether it was coming or going—that is, regardless of how its surface features might be rotated in space. To recognize a turkey as a potential meal, we needed to do the same. This recognition had to be quick, because the wolf or turkey, its own survival instincts programmed by the forces of natural selection, was hardly going to wait for us to do a geometrical analysis by calculation. Today, as a result of natural selection over millions of years, the recognition and differentiation of complex geometries has become innate to humans and most other animals.

In humans, these traits have evolved to a high degree of sophistication. We can pick an old aquaintance from a crowd, simply by recognizing his or her face. Meanwhile, we often complain that photographs of ourselves never seem to look quite right—a consequence of our being more familiar with our reversed mirror image than with our appearance as others and their cameras view us. Although modern computer software is only marginally successful at recognizing a face from a variable perspective, we personally accomplish this task with relative ease dozens or hundreds of times each day.[3] We are masters of recognition of even subtle nuances in morphology.

Biological morphology, however, is not always that complex. Some of the most marvelous wonders of nature are those organic shapes that conform most closely to classic Euclidian geometries. Spheres abound in various fruits, nuts, seeds, tumbleweeds, and even rodent feces. Most eggs are ellipsoidal, and many mushrooms are hemispherical. Fascinating geometries abound in the simplest life-forms; Figure 11.7 shows some of the cones, spirals, helices, and conical helices one finds in the shells of monovalve mollusks. More-developed animals also build structures that may have highly regular geometries: rice birds build nearly spherical nests, African bug-a-bugs (ground termites) build nearly conical mounds, and honeycombs usually have an internal structure consisting of a web of hexagons.

These and similar examples suggest that geometry is not just a formal system of abstract thought. Both the physical and biological worlds are fundamentally geometrical in their workings, and we, as humans, have

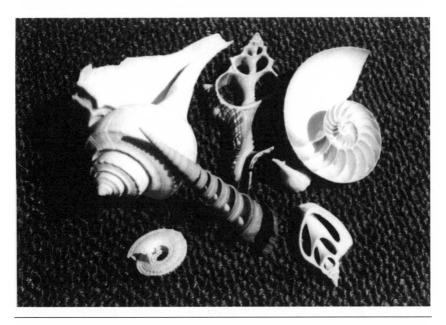

Figure 11.7. The shells of monovalve mollusks and other simple life-forms often display highly regular geometries.

evolved with an innate ability to perceive and even manipulate such natural geometries. Doing so is easy for us, even when such geometries are complex. We and our ancestors have been doing it for millions of years.

What *is* difficult for us is this: figuring out how to describe lines, surfaces, and solids by using less information than the shapes themselves contain. An ellipse, for instance, has an infinite number of points. It is nevertheless possible, as we have seen, to fully describe a specific ellipse by specifying just two numbers: the lengths of its major and minor axes (or alternatively, its major axis and eccentricity). Shrinking an infinite set of coordinates to just two numbers takes quite an intellectual effort, particularly for the first person who figures out how to do so. We humans seem to be poorly equipped to engage in such extremes of reductionism, for the simple reason that our survival in the prehistoric rain forests just didn't depend on this kind of skill.

Given just a few observations, on the other hand, we find it quite natural to proceed in the opposite direction and generalize to the many; we don't need to see every snake, for instance, to know a snake when we see one. This inductive process, once essential to the survival of our species, is programmed into each and every one of us. Generalizations of this type are the essence of science, or at least of protoscience. No matter that we're sometimes wrong, perhaps mistaking a curved fallen branch for a snake, for it does not threaten our survival if we avoid a branch when we think we're avoiding a snake. Inherent in all scientific inquiry is the fact that its generalizations are sometimes wrong. Our thinking apparatus permits us to err, and overall to err a bit more on the side of safety.

Our innate approach to geometry, therefore, is empirical and protoscientific, proceeding from individual observations to somewhat shakier generalizations. We see complexity around us, and we strive to organize it intellectually. Yet our own brains have been wired by natural selection to sacrifice certainty in favor of quickness of conclusion and reaction. Although the intellectual circuitry that governs our thought processes still remains largely a mystery, the result is apparent in all walks of society: we humans jump to erroneous generalizations, we appeal to "common sense" in attempts to elevate subjective nonsense to an intellectual status, and we react emotionally to some of the subtlest of perceived threats. We do these things because our thinking is still governed by Stone Age genes. The limitations on our scientific understanding of the structure of the universe may ultimately lie not in the cosmos, but within our own heads.

The geometry of the mathematicians reverses the direction of this intellectual process, by beginning with abstract axioms and postulates and seeking to derive the properties of classes of shapes, which in turn can be reduced further to specific instances like cycloids and helices. In practice, such axioms and postulates yield comprehensive descriptions only for relatively simple shapes: conic sections, regular polyhedrons, and so on. One does not apply Euclid's postulates to the morphology of snakes and elephants, nor does it seem that it would be productive to do so.

And thus a great gap remains. The universe presents us with marvelously complex geometries, which the methods of empirical scientific inquiry permit us to generalize but poorly. Meanwhile, mathematicians offer us a structure of axioms and theorems that allows us to derive relatively few specific and highly regular geometrical configurations. Yet we can't lightly dismiss the formal thinking of the mathematicians as mere intellectual amusement, for we actually *observe* examples of their derived geometrical conclusions in the physical and biological world! This suggests that geometrical theorems *are* somehow bound up in nature, and that we may gain an understanding of things previously unseen through further extensions of such geometrical reasoning.

A Brief History of the Atom

Cleave an apple in half, split one of the halves in two, then repeat this process again and again. Is there a limit to how far we can go, even with an infinitely sharp knife? The Greek philosopher Democritus (460–370 B.C.E.) thought there was indeed a limit. His logic ran as follows: Because two pieces of matter cannot occupy the same space at the same time, a knife must slice only through space that particles of the apple do not occupy. Therefore, slicing an apple an infinite number of times would require that the apple consisted of nothing but empty space. Empty space, however, obviously does not have the physical properties of an apple. Therefore, there must be some minimum portion of an apple that cannot be sliced further and would repel the infinitely sharp knife. Democritus named this smallest possible piece of matter the *atom.*

We should note that Democritus's logic was abstract, and that his speculations were not a theory in the modern sense of the word, as they lacked the essential requirement of falsifiability. Until quite recently, there existed no technical means to test Democritus's assumptions or conclusion. In fact, as we shall see in chapter 12, even today it remains an open question whether two pieces of matter can occupy the same space at the same time.

Around 1803, John Dalton formulated the law of definite proportions to explain why it always took two times as much gaseous hydrogen as oxygen to produce water. According to Dalton, water is a compound of discrete hydrogen and oxygen atoms; at a basic structural level, two atoms of hydrogen link up with one atom of oxygen to create one molecule of water (H_2O). Other chemists seized on this idea of definite proportions and began weighing everything they combined or reduced to its constituents. Every chemical reaction they studied did indeed seem to follow the same scheme: an integer quantity of one element combining with integer quantities of other elements to produce a compound with new chemical properties. By the mid-1800s, every practicing chemist had added the word "atom" to his scientific vocabulary. Yet as late as 1900, most scientists doubted that these atoms "really" existed; they viewed them, instead, as no more than artificial constructs made computation predictions of the outcomes of chemical reactions possible.

In the mid-nineteenth century, Ludwig Boltzmann proposed a theory that explained the pressure and temperature of gases as necessary statistical consequences of the motion of their tiniest particles. Boltzmann's "kinetic theory" nevertheless fell short of providing direct evidence for the existence of molecules and atoms, which are individually too small to see by any conventional means. It was only around the turn of the century that experimental physicists began to probe the fundamental nature of matter with particle beams, and they concluded that all matter is mostly empty space, punctuated by small pockets of positive electrical charge surrounded by lower-density regions of negative charge. These tiny bundles of interacting electrical charges seemed to correspond to Boltzmann's atoms, as well as to Democritus's atoms of a few millennia earlier. Atoms indeed seemed to exist.

This empirical confirmation of atoms let naturally to the question of their finer structure. The obvious analogy was the solar system. Begin with a positively charged nucleus, set negatively charged particles orbiting this nucleus under electrical rather than gravitational forces, and the result is a miniature solar system, composed mostly of empty space. Niels Bohr was awarded the Nobel Prize in 1922 for developing a quantitative theory of the atom based on this solar-system analogy.

Bohr's theory, however, had a major flaw. The atom cannot be a solar system governed by an analog of Kepler's laws, because electrically charged particles always radiate electromagnetic energy when they travel in curved paths. If an electron speeds in a tight orbit around a positively charged nucleus, it ought to radiate energy at a high rate. As it loses its energy in this manner, an orbiting electron ought to spiral into the atomic nucleus. Combining Bohr's theory with the theory of electromagnetism compels us to conclude that the universe, and everything in it including ourselves, should survive but a few days before collapsing into a dot matrix of atomic nuclei.

There is, however, a way out of this dilemma. Louis de Broglie, as we saw in the last chapter, showed that all matter has wave properties. Electrons bound in atoms have wavelengths of about 10^{-10} meters, which is about the same order of magnitude as the diameter of the atom itself. Thus, because of its wave-particle duality, an electron in an atom cannot be absorbed into the nucleus for the simple reason that it can't fit there! Only in unusual circumstances will a nucleus actually capture one of its orbiting electrons.

Yet we strive for some mental imagery to help us understand how the principle of wave-particle duality might explain the structure of the atom. At the left of Figure 11.8, we see three standing waves of finite length. Such standing-wave patterns are produced in the strings of musical instruments,

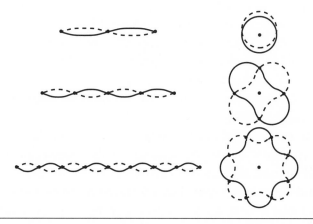

Figure 11.8. An electron orbiting an atomic nucleus can be treated mathematically as a standing wave that wraps back on itself.

as well as in larger structures. (The reader who has seen videos of the 1940 collapse of the Tacoma Narrows suspension bridge will recognize the top diagram as the standing-wave pattern that led to this famous bridge failure.) The essential characteristic of a standing wave is that the finite length of the string disallows all waves except those where an integer number of half-wavelengths fits into the available space. The allowable wave patterns are called the *normal modes* of vibration of the string. A string of finite length either does not vibrate at all or else vibrates in one or more of its normal modes.

Suppose that such a string is wrapped into a closed loop, as shown in the right-hand sketches in Figure 11.8. Then, the normal modes are standing wave-patterns where an integer number of half-wavelengths fits into the circumference. This is our modern imagery of the atom. An electron cannot follow a closed orbit unless its wave closes upon itself geometrically. Placed at some radial position where this is not the case, an electron will give up energy (or else absorb energy, if it is available) and will settle into one or another standing-wave pattern. This standing-wave model explains why orbiting electrons don't spiral into the nucleus. On the scale of the atom, an electron's wave properties dominate its particle properties. An electron cannot collapse into a space smaller than its wavelength, simply because its physical size can't be smaller than its wavelength. Accordingly, an electron's orbit around a nucleus can be no closer than the smallest circumferential path that will accommodate the electron's wavelength.

Electrons do, however, radiate electromagnetic energy, and if this were not so, we would live in a very dark universe. Rather than radiating continuously, an electron gives up energy only when it makes the transition from a higher-mode standing-wave pattern to a simpler pattern within the atom. This energy emission takes place in a discrete quantum jump, and the resulting electromagnetic wave packet, which is identical to a photon, has a frequency $f = E/h$. For this and other reasons, the theory that relates the wave characteristics of electrons to packets of electromagnetic waves is referred to as *quantum theory*.

Yet we should not delude ourselves into thinking that the sketches in Figure 11.8 describe the visual appearance of an atom. Such sketches can be no more than an analogy, one of many that scientists use to comfort themselves as they grapple with the more abstract mathematical representations of quantum theory. We humans grasp such geometrical representations much more easily than we can visualize the meaning of partial differential equations or sets of algebraic matrices. As for what an atom really looks like, the question has no meaning, for atoms themselves are much smaller than the wavelength of any light we could use to illuminate them individually. To some scientists, this inability to observe atoms in any conventional sense suggests that we still can't be sure that they exist apart from the logical structure that defines them.[4]

From Subatomic to Cosmic Structures

Assuming that atoms do exist, however, we can ask a long list of questions about their internal structures. We use two approaches in our inquiries: (1) deriving the implications of our current mathematical theories as applied to new situations, and (2) conducting experiments using ever more sophisticated scientific apparatus. We structure our experiments, some costing millions of dollars, to focus on those questions to which the theorists have already given tentative answers. We focus our mathematical theory-building on those situations in which the experimentalists have found unexpected results. It is in this manner that scientific understanding progresses, in an endless loop of intellectual cross-fertilization between theorists and experimentalists.

The atomic nucleus was "discovered" in the sense that atoms behave in experiments in a way that indicates they have a tiny center of highly localized mass and positive electrical charge. Using an analogy again, if an atom were enlarged to the size of a major league baseball stadium, the nucleus would be "smaller" than the bag marking second base. Everything else would be mostly empty space. If this is correct, then what we perceive as solid matter is an illusion. Everything around us, including our own bodies, is mostly a vacuum.

Of course, this idea raises a whole series of additional questions. How, for instance, do we succeed in simple tasks like driving a nail with a hammer, if the hammer and nail are both empty space? Why doesn't the hammer pass right through the nail? Here, other aspects of the theories of atomic structure and electromagnetism come to the rescue, and we answer on theoretical grounds that the electron clouds in the nail repel the electron clouds in the hammer with a tremendous force that we cannot overcome with our mere human strength. And can matter ever collapse into its own vacuum? Apparently it can, if the electrons swarming about the atomic nucleus are stripped away. In fact, it seems that a sizable fraction of the matter in the universe exists in some form of collapse: both in common stars like our sun, and in other more-compact stars like white dwarfs, neutron stars, and black holes.

One of the deepest scientific mysteries throughout human history has been the question of what makes the sun shine. The ancient Egyptians had their sun-god, Ra, and many other early civilizations had similar polytheistic explanations. The Judeo-Christian view is that the sun is but one of many mysteries of creation of a single God. Monotheism is often the first step on the road to scientific understanding, for it anchors all phenomena in a unified and rational creation. In this context, the most obvious explanation seemed to be that the sun is a fire, essentially the same as a campfire but much more intense. Given, however, that all fires eventually burn themselves out, a new question immediately arose: When will the sun die out?

By the mid-nineteenth century, the science of thermodynamics had come into its own and demonstrated its predictive validity through an impressive range of applications in power engineering. For the first time, it became possible to explore, mathematically, the proposition that the sun is a fire. If one postulates that the sun's fuel is coal, however, it turns out that the laws of themodyamics predict a total solar lifetime of no more than about two thousand years. We can postulate alternative fuels, but no matter what chemical hypothesis we propose, we cannot stretch the sun's lifetime by more than another thousand years or so. In other words, either the sun is much younger than human history, or else it is not sustained by a chemical fire.

Here the matter stood until the 1930s. At that time, theory and experiments were beginning to concur that atomic nuclei and subatomic particles could take part in incredibly energetic processes. One such process, *thermonuclear fusion*, can be initiated only at extremely high temperatures (around 10 million kelvins), far above the temperature where electrons remain in orbit around proton-neutron clusters. Under these extreme conditions, atomic nuclei are no longer isolated from each other by surrounding regions of vaccum; instead, nuclei can interact directly. In doing so, fast-moving nuclei of simple elements like hydrogen and helium will fuse to form more-complex nuclei, releasing energy in the process. In the sun, this outward flow of energy counteracts the inward pull of gravity that would otherwise collapse the nuclear matter into a very small space. According to this mathematical model, we can breathe a sigh of relief about the sun's lifetime. A star like the sun can be expected to be relatively stable in its energy production for about ten billion years. Today our own sun is barely middle-aged.

The structure of a star is dynamic rather than static; it reflects a balance between inward gravitational attraction and outward radiation pressure. When a massive star eventually expels most of its nuclear energy, its outward radiation pressure declines, and the star collapses into a white dwarf, typically only a few hundred miles across, or even into a neutron star, only a few miles in diameter. A neutron star is invisible to the eye, not just because of its small size, but because it radiates at wavelengths too short to be visible. Neutron stars can be observed only through the X-rays they emit.

In a neutron star, there is virtually no space between the particles. The original star's atomic nuclei have all clumped together into a single gigantic super-nucleus whose density boggles the imagination. If it were possible to scoop up a spoonful of a neutron star, it would weigh thousands of tons. Yet even the neutron star has a structure. It is composed of uncountable billions of protons and neutrons that attract each other over short ranges, yet repel each other over still shorter ranges. It is this repulsive aspect of the nuclear force that allows the neutron star to assume a stable size in spite of its own tremendous gravity.

Yet the story is still not over, for we can ask yet another question: What if a neutron star was so massive that its gravity was strong enough to overcome even the nuclear repulsion between nucleons? Now there would be no physical force remaining to counteract a complete gravitational collapse. A complete collapse into what? Into nothing at all; a point; a singularity in space. This is the so-called *black hole*, in which physical matter disappears from the observable universe.

It is not my intention here to delve into all of the fascinating cosmological implications of stellar evolution as we know it; there are many excellent sources available to the reader.[5] Rather, my point is this: the properties of the very largest stars and galaxies of the universe are intimately related to with the properties of the very tiniest of subatomic particles. We as human observers occupy a scale between these extremes. And what are we made of? By all the evidence we have today, we are made of stardust, the remnants of those burnt-out stars that have not disappeared into the black holes of the universe.

CHAPTER 12

THE REAL AND CONJECTURED UNIVERSE

In the twenty-five hundred years since Pythagoras drew seminal connections between scientific inquiry and mathematics, the two disciplines have continuously cross-fertilized each other. Today, science and mathematics have become so intertwined that many college students do not even view them as distinct fields of study. Yet distinct they are, at much deeper levels than any semantic difference. To render informed assessment of what is real in the universe versus what is merely conjectured, we need to begin by examining the fundamental epistemological differences between mathematics and science.

Each branch of mathematics is an axiomatic system that seeks truths that are absolute within that system. The mode of mathematical inquiry is the formal proof, a chain of logic that proceeds deductively from general axioms and postulates toward more specific conclusions. It is on the basis of such formal proofs that we know, for instance, that the sum of the angles of a plane triangle is always π radians, and that a quadratic equation always has two solutions. Formal proofs can also establish nonexistence or nonconstructibility: that it is impossible, for instance, to "square" the circle, to trisect an angle with straight edge and compass, or to find integers that satisfy the relationship $a^n + b^n = c^n$ for any value of n greater than 2.[1]

The foundations of mathematics lie not in computation, but rather in proof, for it makes no sense to compute "answers" to specific cases until we are first assured that such solutions exist. Once the existence of a solution is established, the mathematician can then concentrate on proving that a particular computational method will generate that desired solution. To compute that the equation $x^2 + x = 12$ has roots of $x = 3$ and $x = -4$ is not doing mathematics in its fundamental sense; such a calculation is rather a process of following the computational rules that are consistent with the established theorems, while taking care not to violate any other proven theorems along the way. Computations are not necessarily easier than proofs, and we can go wrong in many ways: miscalculating through either ignorance or misunderstanding of a theorem, or by misapplying a theorem to circumstances outside the bounds of its axiomatic basis. Except for those mistakes where we've simply written the wrong number, computational errors can always be traced to a violation of some proven mathematical theorem.

It's true that proofs and theorems are far from our minds when most of us calculate. We usually think of mathematics in terms of getting specific answers. If we need to know how much concrete to order to pour a driveway, or want our accountant to compute the lowest possible tax we are legally required to pay, we assume that mathematics will come to our rescue, with little thought about its infrastructure of axioms, proofs, and theorems. Yet, just as any traveler is dependent on the integrity of his maps, every computation we perform rises from a foundation of formal proof.

Science, in contrast, rests on an empirical rather than an axiomatic foundation. Nature presents us with a wild assortment of events, and even everyday life would be impossible if we didn't learn to cluster such events into categories and patterns. (The sky is growing dark, and we hear distant thunder? Then surely it's going to rain soon.) Our view of reality is rooted in what we can observe, and our physical laws are based on replicable patterns in such observations. Scientific theories seek to validate groups of such laws, by hypothesizing overriding principles that they seem to have in common. (The oceans have tides, and the planets orbit the sun in ellipses? Then perhaps all massive objects give rise to a force of gravitational attraction.) Science always proceeds from the specific to the increasingly general: from the observed events themselves, to classes of such events, to natural laws that seem to account for such classes of events, to theories that seek to explain the natural laws. Scientific inquiry is an intellectual pyramid, and it proceeds inductively, precisely opposite to the direction of axiomatic mathematical inquiry.

Because there exists no a priori axiomatic basis to science, scientists cannot "prove" their results the way mathematicians can. In fact, the entire process of scientific inquiry may be viewed as an effort to discover such an axiomatic basis, from which all natural events would be necessary consequences. Scientists delight in finding pockets of ignorance about nature's ways, for where we understand least, we have the most to learn. We critically examine some puzzling pattern of natural events; then we take a creative leap and venture a hypothesis. We do not confuse such a hypothesis with "truth." Instead, we use this hypothesis to predict the outcomes of future events we have yet to observe. We may go so far as to create the conditions that will stimulate such events; this process we refer to as "experimentation." And if such new events indeed have outcomes consistent with our hypothesis? Then our hypothesis is supported. It is not, however, proved.[2] To prove a hypothesis in any absolute sense, we would need to observe all similar events in the entire universe for all time.

Instead, if the results of scientific predictions do conform to a hypothesis, scientists say that the hypothesis has been *validated*. Given an increasing number of such validations by different scientists in different settings, a hypothesis gradually gains acceptance within the scientific community and is elevated to the status of a physical law or theory. Absolute truth, however, still remains an unattainable ideal.

In the early 1980s, for instance, Luis Alvarez and his coworkers advanced the hypothesis that the puzzling extinction of the dinosaurs some 65 million years ago resulted from the impact of a large asteroid.[3] The initial evidence was meager; it amounted to an observed high concentration of the rare element iridium in geological samples taken from the narrow boundary between the Cretaceous and Tertiary periods of prehistory, which coincides with the disappearance of dinosaurs from the fossil records. Although iridium is rare on Earth, it is common in metallic meteorites. Initially, the scientific community was skeptical about the fundamental observations, and many scientists set out to contradict Alvarez's notions by analyzing additional geological samples from around the globe. In each study, however, they found further confirmation that the so-called K-T boundary indeed has an elevated concentration of iridium everywhere on the planet.

Alvarez hypothesized that the asteroid impact blasted a large quantity of particulate material into the high reaches of the atmosphere, where it shielded Earth's surface from the sun for several years until the iridium-laden dust settled to the ground around the globe and over the bones of the last of the dinosaurs. Confirming that this iridium layer is indeed found around the globe, however, does not prove the hypothesis. Scientists clamored for the "smoking gun": the impact crater. To fit the hypothesis, calculations tell us that this crater must so be big, around 250 km in diameter, that surely some evidence of it would remain today, even after 65 million years. And indeed, it would seem that an underwater crater (Chicxulub) recently discovered off the southwestern shore of the Yucatan Peninsula of Mexico has the right age if not quite the right diameter (about 170 km).

Although scientific efforts continue to consolidate the evidence, never will it be proved definitively that the Chicxulub impact was responsible for the iridium layer around the globe, or that this event triggered the demise of the dinosaurs. Scientific explanations always remain speculative. Although corroborating evidence helps validate a hypothesis or theory, no theory can possibly be proved beyond all shadow of doubt. The best we can say about Alvarez's theory of dinosaur extinction is that it is supported by all relevant observations made to date, and that no one has proposed a compelling competing theory.

Scientific hypotheses and theories can, however, be *disproved*. This is perhaps the quickest way for a young scientist to build a reputation: take someone's theory, predict its consequences, then document one's own observations that contradict these predictions. If real events contradict a theory's predictions, and the observations stand the test of independent scrutiny, then that theory is indeed disproved. The theory of natural selection, for instance, would collapse if someone found a few organisms whose DNA molecules spiral counterclockwise rather than clockwise. In fact, some of the most important experiments in the history of science have been those that have pulled down the edifices of prior theory in this manner.[4]

Clearly, the criteria for truth are different in science and mathematics. Although mathematics and science are equally capable of *disproving* a hypothesis, it is only mathematics that proves. Mathematicians can be confident of their abstract truths; scientists can at best be tentative about truth in the natural universe. In mathematics, truth can be discovered through an individual effort that is consistent with to formal rules of logic; the result can then be announced to the rest of the mathematical community for concurrence. In science, however, an individual effort does not go very far. An empirical observation cannot be accepted by the scientific community unless independent researchers physically replicate and validate that observation, which often may not be easy to reproduce, and which inevitably raises a whole host of practical issues relating to physical hardware and instrumentation. Science, therefore, is inherently a social and consensus-building road to truth.

As a corollary, scientific inquiry ignores many phenomena that are undoubtedly quite real, but rare in the course of human events. Given a report of an instance of ball lightning plummeting into a rain barrel and boiling away the water, a scientist simply cannot proceed. There is no way to replicate the event, theorize without a large sample of prior observations, or predict future ball-lightning events in a way that would enable someone to put cameras and instruments in place to record them. The scientist therefore shrugs and moves on to other more tractable problems, leaving the issue of ball lightning to await an expanded body of validated observation.

Scientists use mathematics much the same way as the rest of the population: for calculating. Only in some of the more esoteric niches of mathematical physics do scientists go about proving new theorems that no mathematician has ever thought of. For the most part, we grab onto those mathematical theorems that are already available, and we compute accordingly. And what do we compute? In most cases, we compute the future. We apply mathematical logic to our theories to predict the outcomes of events that have not yet been observed. In 1915, for instance, Einstein used mathematical logic to predict that light from a distant star would be deflected by a strong gravitational field. Yet the strongest gravitational field in the solar system, that near our sun, coincides with the region where sunlight obliterates the images of all stars. To test Einstein's prediction meant waiting for a total solar eclipse. The next one occurred in 1918, and a scientific expedition to the South Pacific indeed observed that a star that was actually behind the sun at that time appeared displaced to a position next to of the eclipse. Although this predicted observation didn't *prove* the truth of the theory of general relativity, it did generate a great deal of support for that theory within the scientific community. And why? Because the theory had successfully predicted a phenomenon never seen before with human eyes.

Scientists depend on mathematical logic to generate their theoretical predictions. It would be comforting to be assured that, given a theoretical proposition expressed in mathematical language, any and all logically consistent predictions of that theory could be computed. Unfortunately, it turns out that even some purely mathematical truths are not demonstrable.

This essential incompleteness of mathematical systems was proved in 1931 by the mathematician and logician Kurt Goedel.[5] Goedel's incompleteness theorem may be stated as follows: *Every formal logical system contains truths that cannot be proved within that system.* Clearly, calculations cannot be based on any unproven mathematical statement, even if it does happen to be true. Therefore, even if scientists should someday arrive at an ultimate "theory of everything," it will still be impossible to examine each and every logical consequence. It seems that some truths, both mathematical and scientific, will forever remain elusive.

Physical Reality

Science, as we have seen, is always grounded in observations, and its theories are validated or falsified through additional observations. But are scientists capable of observing all there is? To most twentieth-century scientists (particularly those grounded in quantum theory), this is a tautological question that deserves a tautological answer: Of course we can observe all that is real, we say, for reality itself is defined only by that which is observable. If an event occurs that we cannot observe, then what possible falsifiable theory could claim that that event ever happened? Cold nuclear fusion, N-rays, and the existence of the ether, among other examples, have failed this essential requirement of observability and have rightly been relegated to the trash heap of science. We assign the status of "reality" only to that which we observe, and we observe only that which is real, in a vicious but hopefully expanding circle of knowledge.

Yet such a response is far from complete, for scientists include under the umbrella of "observation" any event that *can be reduced* to one that can be detected by human senses. Such reductions may take place through intervening physical instruments, as in looking at bacteria under a microscope, or through mathematics, as in "observing" the moon's gravitational field by analyzing ocean tides on Earth, or through a combination of both approaches. Indeed, the successes of all modern science, and its spin-off technologies, rest upon such extensions of our feeble human senses. As we use physical instruments and mathematics to extend our senses in this manner, we also extend our concept of physical reality.

As an example, let's consider radio waves. No one ever sees a radio wave, feels a radio wave through physiological tactile senses, or hears a radio wave directly (if indeed we did, we'd all quickly be driven crazy). We do, however, "observe" radio waves indirectly, by intercepting them with a receiving antenna, converting them to electrical signals, processing these signals (e.g., demodulating them) with electronic circuitry, then ultimately converting them to light on a screen or sound from a speaker. We observe, not the radio waves themselves, but the *effects* of these waves, confident that Jay Leno's jokes as we hear them are indeed the same jokes that were encoded into the electromagnetic waves at their point of origin. Although our evidence remains quite indirect, none of us doubts the reality of such radio waves. We take as a given that Jay Leno's real performance has been encoded into radio waves by a set of instruments at the transmitting station complementary to those we use at our receiving end to decode those waves. We conclude, quite reasonably, that the reality of radio waves is as well established as the reality of Jay Leno himself.

The concept of "observation" in our modern world has progressed far beyond the limits Aristotle or even Galileo would have imposed on it. Today, we observe a great many events that go beyond what we can see, hear, feel, smell, and taste firsthand. This expanded concept of observation has succeeded because it has served our purposes well — purposes both utilitarian and inquisitive. Many of us have observed germs that are too small to see, distant galaxies that are too dim to see, underwater scenes that are too remote to see, and internal bodily events (e.g., embryo growth) that are too inaccessible to see. Yet we routinely include these and a host of other indirect observations in our expanded concept of reality. We do this as individuals, and we also do this as scientists.

The Discovery of the Neutrino

The issue of what we mean by physical reality is not always easy to decide. Let's explore the history of scientific thought regarding a more obscure example, an elusive subatomic particle known as the *neutrino*. Our question is whether neutrinos really exist.

By the year 1915, studies of beta radioactivity had revealed that certain unstable atomic nuclei emit electrons spontaneously, and in doing so, they transform themselves into nuclei of a different chemical element. These ejected electrons (β-particles) were found to have a whole range of possible velocities, from nearly zero to almost the speed of light, even when the emitting nuclei were in all measurable respects identical. This observation itself was disconcerting: identical initial conditions leading to a wide range of observed outcomes. More unnerving to the physicists, however, was that the β-decay process seemed to violate a well-established physical law, *the law of conservation of energy*, which states that no physical process ever creates or destroys energy, and that energy can only be changed from one form to another. If identical nuclei emit β-particles of highly variable energies yet settle into virtually identical energy states after these ejections, then any energy-accounting arithmetic would seem to be rendered invalid.

This was a profound dilemma to the physicists of the 1920s. Most of their advances of the previous century had been built upon the paradigm of conservation of energy, or at least were consistent with that principle. To abandon energy conservation would demolish the foundations of all of thermodynamics, most of mechanics, and much of electromagnetic-field theory. Accordingly, these observations of the variable-energy products of β-decay were checked and rechecked again and again through many experiments, but always with the same result: the energies just didn't add up.

Obviously, the phenomenon of β-decay is far removed from human sensory experiences: nuclei cannot be seen with the eye, nor can electrons, nor can particle energies be sensed except indirectly through intermediate detectors whose outputs need to be interpreted through prior physical theory and its attendant mathematics. Researchers appropriately questioned every step through this epistemological quagmire, yet could not significantly alter the final, albeit indirect, "observations." The energies still didn't add up, regardless of how well the experiments were conducted or analyzed.

By 1929, some of the greatest physicists of the time, Niels Bohr among them, were ready to sacrifice the law of energy conservation in favor of these indirect observations. Bohr himself stated: "We have no argument for upholding [the law of energy conservation] in the case of β ray disintegrations. The features of atomic stability responsible for the existence and properties of atomic nuclei may force us to renounce the very idea of energy balance."[6] Other physicists took a wait-and-see attitude, and in December 1930 Wolfgang Pauli wrote to his colleagues,

> I have come upon a desperate way out. To wit, the possibility that there could exist in the nucleus electrically neutral particles . . . [neutrinos]. The mass of the neutrinos should be not larger than 0.01 times the proton mass—the continuous β spectrum would then become understandable from the assumption that in β decay

a neutrino is emitted along with the electron, in such a way that the sum of the energies of the electron and the neutrino is constant.

I admit that my way out may not seem very probable *a priori* since one would probably have seen the neutrinos a long time ago if they exist. But only he who dares wins. One must therefore discuss seriously every road to salvation. . . . Thus, dear radioactive ones, examine and judge.[7]

A few years later, in 1932, James Chadwick discovered another subatomic particle, the *neutron*, which has a mass slightly larger than the proton's mass and 1,839 times the mass of the electron, but with no electrical charge. Neutrons can be observed only indirectly, by examining how nuclei recoil when they interact with each other and with free particles. Neutrons assure that atomic masses add up mathematically, and it was essentially through such an accounting procedure that the neutron was "discovered." The neutron is usually stable when bound within a nucleus, but when naked it is found to be unstable, disintegrating (in an average of about seventeen minutes) into a proton and an electron. This phenomenon of neutron disintegration turns out to be another example where the energies don't add up, unless one postulates that a *neutrino* (actually, an antineutrino) is also emitted whenever a neutron blows itself apart. Now, physicists were confronted with two quite distinct phenomena where energy conservation was violated unless one postulated the participation of a neutrino. Still, the evidence for the reality of the neutrino remained extremely indirect.

In the following years, new theories were proposed to explain why the sun shines. The only solar theory consistent with documented measurements of sunlight intensity required that the sun emit a great swarm of nearly massless neutrinos along with its electromagnetic radiation. According to this thermonuclear-fusion model, about 10 trillion neutrinos streak through our bodies each second, yet many years must elapse before even one of these highly elusive particles happens to interact with one of our own atomic nuclei. Most of this incredible number of neutrinos pass not only through our bodies, but through the entire planet Earth, unaffected by other physical matter as they speed toward the far reaches of the universe. Nuclear theory requires that the neutrino, if it exists, interacts extremely weakly with ordinary matter.

According to most sources on the subject, the neutrino was officially discovered in Savannah, Georgia, in 1956. Again, this discovery was not an explicit observation, but rather a finding implied by the alteration of a small sample of nuclear processes that should have happened otherwise in the absence of a neutrino. Many physicists were skeptical that these detected anomalies amounted to a "discovery," and today some are still skeptical. As one physicist has written, "I believe there is little reason to be convinced by this experiment that the neutrino exists apart from the theory and experiments that define it."[8]

Despite the ambiguities in the evidence for the existence of the neutrino, specially designed neutrino-detection devices were being built by the 1970s. Because the neutrino interacts so weakly with other matter, such detectors need to be very large, often holding hundreds of thousands of gallons of liquid. (One liquid used is cleaning fluid, each molecule of which contains several chlorine atoms.) To shield these huge detectors from contaminating cosmic rays, they and their accompanying sensors and electronics are placed deep underground, usually in abandoned mines.

And what do these giant underground detectors allow us to "observe"? Usually a few events per month, and a few handfuls of events per year, that are sufficiently anomalous to be attributed to neutrinos—hardly enough for the nonscientist to jump in excitement and declare that this is a *phenomenon*! To most of the general public, neutrino detection is not very exciting science: waiting for weeks for an event or two that may help validate the theoretical expectation that there are 1×10^{14} neutrinos passing through each square meter of Earth's surface each second, only a handful of which actually interact with anything.

To the scientist, however, time is relative and patience a virtue. Data on neutrinos have now been collected for more than two decades. And it turns out that these data pose a new puzzle, for the "observed" neutrino-flux density is too low to account for the theoretical models of the thermonuclear processes in our sun. Some one- to two-thirds of the expected solar neutrinos are missing, and we don't have a good reason why. Are we wrong about our theory of how the sun shines, or are we wrong about how to detect neutrinos? Maybe, in fact, we are wrong about the existence of neutrinos themselves.

In the early morning of February 23, 1987, something curious happened: neutrino detectors in Ohio and Japan simultaneously detected a "shower" of nineteen neutrinos.[9] A few hours later, skywatchers in the Southern Hemisphere observed an unusual event in the heavens that was visible even to the naked eye: a star within a neighboring galaxy called the Large Magellanic Cloud exploded into a supernova. This explosion occurred some 170,000 light-years from Earth, which is to say that we witnessed the event 170,000 years after it happened. The close correlation between the timing of the neutrino detections and the arrival of the visible light is compelling evidence that the detections were not mere glitches in the instruments.

Yet supernovas are rare; the 1987 event was the first one visible from Earth in the last four centuries. Our supernova measurements are thus based on a sample of just one, and we are in for a long wait if we wish to corroborate these observations by studying another supernova. Today, the more cautious scientists still demur about declaring the neutrino a reality.[10] And even if the naysayers should concede that neutrinos do exist, additional questions remain to be explored: Do neutrinos have mass? And where are the missing solar neutrinos that are predicted by theory?

Most scientists believe that their business is to discover truth rather than to construct truth. If, however, we agree that the neutrino has been discovered, we also have to agree that scientific discoveries can be highly indirect, and more dependent on mathematical arguments than on sensory experience. As scientists probe deeper into the unanswered questions of the universe, studying ever more subtle phenomena, we can expect to read of an increasing number of "discoveries" whose relationships to objective reality are tenuous at best. Accordingly, we increase our collective risk of confusing the essence of the physical universe with its mathematical messengers.

The Limits of Reality

As science probes the objective realities of the universe, over and over again it finds fundamental limits on what an observer can perceive or predict.[11] We have seen that relativity theory, while restricting masses to speeds lower than the speed of light, also establishes event horizons beyond which we cannot probe. The first law of thermodynamics denies the possibility of events that would result in a net gain in energy, but in building our theories and instruments in accordance with this law, we forever preclude the possibility of detecting a contradicting event. The quantum uncertainty principle prohibits us from increasing the precision of a measurement unless we simultaneously increase our uncertainty about some related quantity, and it decrees that the very act of measurement always has some effect on the event we seek to measure. More recently, chaos theory has established the impossibility of the complete prediction of future configurations of complex macroscopic systems. All of these restrictions, of course, can be expressed in formal mathematical language. Their combined effect is to reduce our options for making observations, thereby limiting the scope of what we can legitimately claim to be physical reality.

Meanwhile, as we probe the universe at levels that approach the self-limitations of our theories, we continue to confront new and ever-more-profound questions. Perhaps the deepest of these is the issue of the *missing mass*. Most galaxies, like our own Milky Way, rotate about their centers. Stars near the outside of a galaxy are always observed to orbit the center at higher speeds than stars near the middle. This is just the opposite of the behavior of our own tiny solar system, where the outer planets travel relatively slowly compared to the inner planets. Why the difference? The current theory of gravity gives but one way out, for it predicts (at least for roughly circular orbits) that the speed of an orbiting body is given by

$$v = \sqrt{\frac{GM}{r}},$$

where G is the universal gravitational constant, r is the radius of the orbit, and M is the total mass contained within that orbit. In our solar system, where virtually all of the mass is concentrated in our single sun, the speeds of the planets are indeed inversely proportional to the square roots of their orbital radii. In a galaxy, however, there are vast numbers of suns, and the outer stars orbit a much larger central mass M than do the inner stars. This *ought* to explain why the outer stars travel about the galactic center faster than the inner stars.

Unfortunately, this explanation stands at odds with the observations. The speeds of the outer stars in most galaxies are too great by a factor of at least three to satisfy this equation. Clearly, we are either wrong in our theory of gravity, or we are wrong in establishing the mass M of a swarm of galactic stars. At present, the consensus is the latter: matter is missing that we don't seem to be able to see. And how much is missing? Around 90%.

There have been many suggestions for the forms this missing matter might take. One of the most intriguing suggestions is this: We know that neutrinos interact very weakly with ordinary matter—so weakly, in fact, that they hover on the edge of our concept of reality. Assuming that neutrinos are real, however, isn't it possible that there could exist still other particles that don't interact with ordinary matter at all, except perhaps through their gravity? Even a star-sized clump of such particles would be invisible, given no mechanism (e.g., electromagnetic waves) by which we could interact with it as observers. Such a form of matter might even occupy the same physical space as Earth, yet be undetectable because it would exert no forces on our protons, electrons, and neutrons.

Yet in such a speculation we have gone beyond the bounds of science, for what is not detectible cannot be considered to be real. At some level—and perhaps this is it—science may be destined to merge once again with philosophy, the discipline it split off from many centuries ago. Ultimately, it may be the mind alone that is capable of grappling with the very deepest questions of the nature of reality, and Aristotle's notions on this will have been at least partially redeemed.

The Mystery of Mathematical Applications

It is always a wonderful experience to confront some puzzle of nature, refer back to some previously established scientific theory, write the appropriate equations, and find that the mathematical solutions indeed account for the observations. I'm also astounded each time NASA succeeds in sending an instrument package to a distant planet, using computers programmed with Newton's laws of motion and gravity. Such circumstantial evidence suggests that the universe as a whole is intrinsically mathematical.

Before we close the book on such a conclusion, however, we need to explore it further. Nobody, it turns out, has ever proved that erroneous mathematical logic is *incapable* of predicting phenomena that are verifiable by experiment. Meanwhile, the history of science abounds with instances where the results of valid mathematical logic have conflicted with reality. In seeking to identify those mathematical formalisms that describe real observations, scientists experiment with mathematics much as they experiment with natural phenomena: they hypothesize, they test, and they evaluate whether the mathematical system they've employed is physically valid.

The actual state of affairs, then, is that mathematics sometimes links up with reality, and sometimes it doesn't. In repeating instances of our successes and abandoning our failures, we drive a mechanism of natural selection that favors the survival of those paradigms that work in the real world. The mathematics that enters our science textbooks and research journals is the mathematics that has been demonstrated to be useful in physical contexts. Other findings of pure mathematics (e.g., Fermat's last theorem) remain beyond the fringes of scientific relevance.

Yet even this cannot be the whole story, for the history of mathematics is replete with instances of theorems and mathematical systems that began as purely intellectual inquiries and only generations or centuries later found practical application by scientists. Riemannian geometry, for instance, was available for Einstein to incorporate into general relativity, and group theory was ready when the nuclear physicists needed it. Why is it that scientists so often have been lucky enough to find that the mathematics they need to develop scientific theory has already been developed?

The answer seems to unfold at several levels, beginning with geometry. Given that all physical events take place in time and space, spatial relationships necessarily enter most scientific theories. Geometry is empirical in its origins, rising out of ancient needs to survey land and to build structures, and its axiomatic basis is grounded in concepts that are at least approximated by physical entities (lines, points, polygons, and circles, for instance). Because geometry arose as an abstraction from physical reality, the approximate validity of formal geometry for physical description is not all that mysterious. The practical successes of formal geometry, moreover, became a force that drove scientists to consider whether other more abstract mathematical systems might also be useful in scientific description and theory building.

The second reason that the universe seems so mathematical is that science ignores those branches of mathematics that don't apply. Again, the decision of what math to use rests on empirical validation. For Isaac Newton, there wasn't any preexisting mathematical system that would allow him to solve his theoretical equations of orbital mechanics; he was therefore forced to either give up on his theory or else invent a new branch of mathematics. Newton chose to do the latter, and invented the calculus. Subsequent observational validation of his physical predictions in turn helped validate the calculus itself, attracting the attention of other mathematicians, who put it on a firm axiomatic basis.

In other cases, quantum mechanics for instance, we have more than one choice of appropriate mathematical system. We can cast quantum theory in the mathematical formalism of partial differential equations, or alternatively as a set of matrix equations. These two formalisms not only look different symbolically, but they have different sets of rules for computation. Either way, however, we reassuringly arrive at the same mathematical predictions of energy states and physical probabilities that are consistent with observation. At least for quantum-mechanical computations, we can base our choice of mathematical system on considerations of convenience.

Yet there may be another answer, admittedly a bit more speculative, to the question of why the universe seems intrinsically mathematical. Given that mathematics is a creation of the human mind, all mathematical logic must flow from at least one form of physical reality: the human brain. Eons of biological evolution have structured our brains to solve problems in ways that enhance our chances of survival as a species. An intuitive sense of geometry is essential; without it the members of a hunter-gatherer society will get lost in the forest or be devoured by a predator of a misinterpreted morphology, and thereby be denied the opportunity to pass on their genes. Depending on our individual ancestry, the Stone Age was between two hundred and perhaps four thousand years ago, or only some 8 to 160 human generations. The slow grind of biological evolution does not accomplish major changes in such a short time; in many ways our Stone Age genes are still apparent in modern society.

The fact that it is so natural to think geometrically may stem from this same genetic ancestry. Even when more-abstract mathematics is involved, scientists still try to reduce their mathematical models to pictorial representations, analogous to the cave painting. They routinely create diagrams of frequency distributions, atoms, black holes, and structures in phase space, when in fact none of these geometries are directly observable. And why? Because they know that their readers will immediately relate to the geometry, whereas some may be confused by other forms of symbolic mathematics.

The human brain is homologous with the universe; it is composed of atoms and molecules identical to those in distant galaxies, and it is subject to the same natural laws that govern the motions and chemistry of comets. To the extent that our thought processes are governed by natural laws, it may be that many of the answers we seek in the "external" world already reside within our own biological thinking apparatuses. Our brains, incapable of any activity that the laws of nature prohibit, may be structured with an inherent inclination to invent those mathematical systems that conform to physical reality. If this is the case, then mathematics may apply to reality because this mode of thought is a direct result of nature setting limits on the activities of the human brain. Biologically, we may not be capable of dreaming up mathematical schemes that drastically conflict with physical reality.

Yet this remains conjecture, based on a sample of just one thinking species. It would be interesting to encounter another advanced civilization somewhere in the far reaches of the galaxy. Would such creatures have invented the same mathematics as we earthlings? Would they know about circles, and about π? Would they use a similar algebraic logic? How about differential equations and matrices, probability and statistics, and the calculus of complex variables? And if so, would they use such mathematics in the same practical ways: in building scientific theories and setting standards of practice in engineering and finance? I personally suspect that they would. But then, being human and composed of stardust, maybe I am not allowed by the laws of nature to suspect otherwise. Maybe creatures made of the "missing matter" of the universe would invent completely different analytical systems, unintelligible to us even if there existed a means to communicate across noninteracting matter.

Our universe is a big place, and we'll never come close to observing all of it for all time. Many puzzles confound scientists today, and new mysteries continue to arise. Perhaps mathematics illuminates the workings of nature only in some areas, and the most intractable problems (e.g., the missing-mass problem, or the problem of finding a unified field theory) are of such a nature that they don't conform to any mathematical system we humans are biologically capable of devising. If this turns out to be the case, it will be a big disappointment. At present, however, it's much too soon to concede to any speculative limitation on the applicability of mathematics to science. Before we can definitively evaluate such a premise, we will need at least a few more centuries of intensive intellectual effort in using mathematics to seek answers to the outstanding puzzles of the universe.

APPENDIX A

Formulas for the areas of common shapes

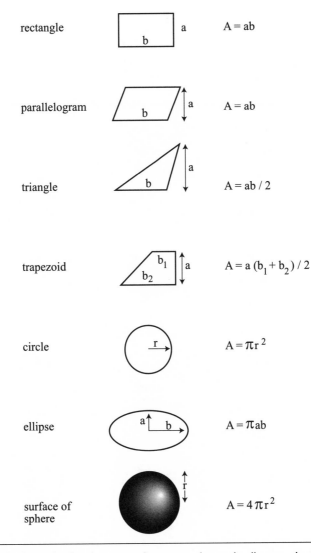

rectangle		$A = ab$
parallelogram		$A = ab$
triangle		$A = ab\,/\,2$
trapezoid		$A = a\,(b_1 + b_2)\,/\,2$
circle		$A = \pi r^2$
ellipse		$A = \pi ab$
surface of sphere		$A = 4\,\pi r^2$

Appendix A. Formulas for the areas of common shapes. In all cases, the formula calls for the product of two linear dimensions.

APPENDIX B

Formulas for the volumes of common solids

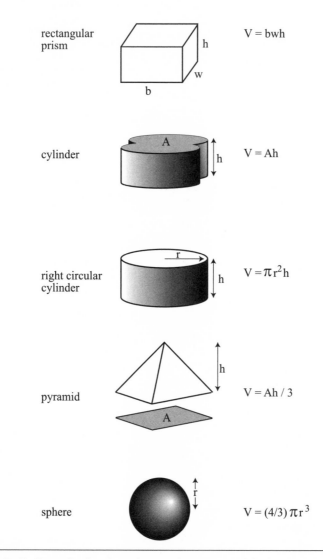

rectangular prism $V = bwh$

cylinder $V = Ah$

right circular cylinder $V = \pi r^2 h$

pyramid $V = Ah / 3$

sphere $V = (4/3) \pi r^3$

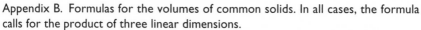

Appendix B. Formulas for the volumes of common solids. In all cases, the formula calls for the product of three linear dimensions.

APPENDIX C

Algebraic equations for the conic sections

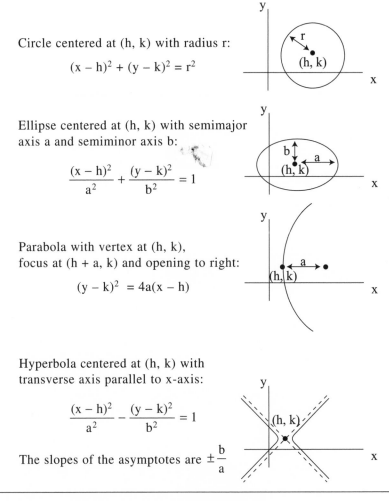

Circle centered at (h, k) with radius r:

$$(x - h)^2 + (y - k)^2 = r^2$$

Ellipse centered at (h, k) with semimajor axis a and semiminor axis b:

$$\frac{(x - h)^2}{a^2} + \frac{(y - k)^2}{b^2} = 1$$

Parabola with vertex at (h, k), focus at (h + a, k) and opening to right:

$$(y - k)^2 = 4a(x - h)$$

Hyperbola centered at (h, k) with transverse axis parallel to x-axis:

$$\frac{(x - h)^2}{a^2} - \frac{(y - k)^2}{b^2} = 1$$

The slopes of the asymptotes are $\pm\dfrac{b}{a}$

Appendix C. Algebraic equations for the conic sections.

NOTES

Chapter 1. The Quest for Pi

1. The value of 3 for the constant circle ratio is given in the Hebrew Old Testament in 1 Kings 7.23 and 2 Chronicles 4.2.

2. For further details, see P. Beckmann, *A History of Pi* (New York: Golem Press, 1971).

3. The requirement of a 10-meter-diameter circle can be confirmed as follows: Given a precision of measurement of ±1 mm, the circle's diameter actually lies somewhere between 9,999 mm and 10,001 mm. This same circle's circumference ought to be 31,415.9 mm, but because we are limited to a precision of ±1 mm, physical measurement can establish the circumference only as something between 31,415 mm and 31,417 mm. This leads to an upper limit on $\pi = C/D$ of

$$\frac{(31{,}417 \text{ mm})}{(9\,999 \text{ mm})} = 3.142\,014$$

and a lower limit of

$$\frac{(31{,}415 \text{ mm})}{(10\,001 \text{ mm})} = 3.141\,186.$$

Thus, even with a 10-meter-diameter circle, the upper and lower limits on π do not quite round off to the same third decimal place. To find π to greater precision using such a measurement approach, we would need an even larger circle.

4. For an account of Archimedes' death, see T. L. Heath, *A History of Greek Mathematics* (Mineola, N.Y.: Dover, 1981). See also E. J. Dijksterhuis, *Archimedes* (Princeton, N.J.: Princeton University Press, 1987).

5. Gregory's original reasoning is not well documented. After Brook Taylor's work on power series (1715), it became possible to show that Gregory's series for π follows from the Taylor series expansion for the inverse tangent function:

$$\tan^{-1} x = x - \frac{x^3}{3} + \frac{x^5}{5} - \frac{x^7}{7} + - \cdots, \quad \text{where } /x/ \leq 1.$$

Setting $x = 1$: $\tan^{-1} x = 45°$, or $\pi/4$ radians. Then,

$$\frac{\pi}{4} = 1 - \frac{1}{3} + \frac{1}{5} - \frac{1}{7} + - \cdots,$$

and multiplication by 4 yields Gregory's series. Taylor's work also explains why Gregory's series converges so slowly, and indeed for any value of $x > 1$, the above series does not converge at all.

Chapter 2. Rollers, Wheels, and Bearings

1. Paul Bahn and John Flenley, *Easter Island, Earth Island* (London: Thames & Hudson, 1992). See also J. Diamond, "Easter's End," *Discover*, Aug. 1995, pp. 62–69.

2. E. L. Newhouse, ed., *The Builders* (Washington, D.C.: National Geographic Society, 1992). See also *This Old Pyramid* (video), prod. P. S. Apsell, NOVA and WGBH Boston, 1993.

3. P. Eschmann et al., *Ball and Roller Bearings: Theory, Design, and Application*, 2d ed. (Ann Arbor: Books on Demand, 1985).

Chapter 3. The Celestial Clock

1. The constellations have changed shape in the last few thousand years, and in ancient times today's North Star (Polaris) was not at the center of rotation of the night sky. We now know, of course, that the stars do not actually move during the course of a single night, and that their apparent motion is a consequence of Earth's rotation on its axis.

2. H. A. Klein, *The World of Measurements* (New York: Simon & Schuster, 1974), chap. 8.

3. The angular resolution of the human eye, typically about two arc-minutes, is primarily a consequence of the coarseness of the rod and cone receptors of the retina. This physiological structure varies little from person to person.

Chapter 4. Mathematics and the Physical World

1. J. Bronowski, *The Ascent of Man* (Boston: Little, Brown, 1976).

2. Some readers may take exception to this statement, in view of modern quantum measurements of nuclear charge, electron spin, photon polarization, and so on, which are reported in the literature as either integers or rational fractions. My response is that even quantum numbers are not measured exactly in the lab, and that even if they were, they would still only represent a rational-fraction multiple of some other measurement (the electronic charge, Planck's constant, etc.).

3. J. N. Wilford, *The Mapmakers* (New York: Vintage, 1982).

Chapter 5. Charting the Planet

1. Arguments for the location of the prime meridian went beyond appeals to national pride, for there must be one line of longitude that defines the change in

the calendar day. Clearly, all trading nations wanted to use the same "date line." There are computational advantages to placing this date line 180° from the prime meridian, and there are practical advantages to running it over the ocean rather than inhabited land areas. Because England happens to be 180° from the broadest north-south expanse of the Pacific Ocean, running the prime meridian through England was a natural multinational choice.

2. A comprehensive historical account can be found in D. Sobel, *Longitude* (New York: Walker, 1995).

Chapter 6. Surface and Space

1. The typical problem is to determine the size of the conductor needed to carry a given current (amperage) while incurring no more than a given voltage drop per unit length. Because voltage drop is inversely proportional to a wire's cross-sectional area and wires are circular, the American Wire Gauge system (AWG) is structured on areas of circles. For more details and tables, the reader can consult the current (U.S.) National Electrical Code, or most electrician's handbooks.

2. If Saturn had the same composition as Earth, it would be around a thousand times as massive as Earth; the fact that it is only about a hundred times as massive is an indication that it is composed mostly of gases, possibly with an interior of liquid hydrogen. The core of planet Earth seems to be molten iron.

Chapter 7. Celestial Orbs

1. This discussion is not intended to disparage Catholicism as a religion, but rather to point out the intellectual pitfalls of attempting to answer scientific questions through authoritative decrees.

2. E. A. Moody, *Truth and Consequences in Medieval Logic* (Greenwood Press, 1976).

3. Nicholas Copernicus, *On the Revolutions*, ed. J. Dobrzycki (Baltimore: Johns Hopkins University Press, 1992).

4. Johannes Kepler, *Epitome of Copernican Astronomy and Harmonies of the World*, trans. C. G. Wallis (Buffalo: Prometheus Books, 1995).

5. A geometrical interpretation has been forthcoming only recently. See M. Bucher, "Kepler's Third Law: Equal Volumes in Equal Times," *Physics Teacher* 36 (4) (Apr. 1998): 212–214. See also M. Bucher and D. Siemens, "Average Distance and Speed in Kepler Motion," *American Journal of Physics* 66 (1998): 88.

6. See Galileo Galilei, *Dialogues concerning Two New Sciences* (Buffalo: Prometheus Books, 1991). For a socially sensitive interpretation of Galileo's conflict with the church, I also recommend Berthold Brecht's play *Life of Galileo*, ed. J. Willet and R. Manheim (New York: Arcade Publishing Co., 1994).

Chapter 8. From Conics to Gravity

1. Carl B. Boyer, "Analytic Geometry: The Discovery of Fermat and Descartes," *Mathematics Teacher* 37 (Mar. 1944): 99–105.

2. Bronowski, *Ascent of Man.*

3. Although various nonorthogonal coordinate systems have been studied by mathematicians, they are all hideously complicated and are generally avoided for practical applications. Orthogonality ensures that the individual coordinates vary independently as one moves in a single coordinate direction (e.g., if we move only horizontally, there is no change in the vertical coordinate, and if we move only radially, there is no change in the angular coordinate). The only physical situations that compel the use of nonorthogonal coordinates are those described by the general theory of relativity, first proposed by Albert Einstein in 1915. In regions of strong gravity, it turns out, no movement is possible that affects only a single coordinate. For a further discussion, see Albert Einstein, *Relativity* (New York: Crown, 1961).

4. F. Graham, "A Hypothetical Ancient Telescope," *Horus* 2(3) (1983): 25–28.

5. The possibility of a hyperbolic lens was noted by Descartes; see his *Discourse on Method, Optics, Geometry, and Meteorology*, ed. Paul J. Olscamp (Indianapolis: Bobbs-Merrill, 1965), 139. This conjecture cannot be proved by analytic geometry alone, for the refractive properties of the glass also play a part in the physical effect.

6. This incident and other aspects of Newton's mathematical, scientific, and political life can be found in V. F. Rickey, "Isaac Newton: Man, Myth, and Mathematics," *College Mathematics Journal* 60 (Nov. 1987): 362–389.

7. See D. Schrader, "The Newton-Leibniz Controversy concerning the Discovery of the Calculus," *Mathematics Teacher* 55 (May 1962): 385–396.

Chapter 9. Oscillations

1. General Congress on Weights and Measures: 13th CGPM (1967), Resolution 1.

2. E. Bruton, *Clocks and Watches* (London: Hamlyn, 1968), chap. 1.

3. Of course, there do exist approximate theories of friction, which can be used to make ballpark predictions of the frictional forces arising in a particular set of circumstances. If we need to know more precise values, however, we have to conduct direct laboratory tests and take measurements, and still be prepared to find that such measurements vary from sample to sample and as environmental conditions change.

4. Huygens described this clock, and other aspects of his contributions to timekeeping theory and practice, in his *Horologium*, first published in 1658.

5. Peter Dirichlet put Fourier's seminal work of 1807 on a firm mathematical footing in 1823, and extended some of this work in 1837.

6. The mathematical procedures and formalisms of Fourier analysis are described in most advanced calculus textbooks. See also D. C. Champeney, *A Handbook of Fourier Theorems* (Cambridge: Cambridge University Press, 1987).

7. One can purchase electronic Fourier analyzers, which will give the amplitudes and periods that make up a periodic waveform. No computations are performed in most of these devices; they operate by physically sensing the waveform's individual harmonic components.

8. Christiaan Huygens, *The Pendulum Clock, or Geometrical Demonstrations concerning the Motion of Pendula as Applied to Clocks*, trans. Richard J. Blackwell (Ames: Iowa State University Press, 1986). For a formal derivation, see note to Table 9.1.

9. It was not uncommon, for instance, for medieval windmills to catch fire during windstorms, due to frictional heating in their gear trains when they rotated too fast. This eventually led to the invention of the governor, which maintained a relatively constant rate of rotation by altering the pitch of a mill's vanes in response to variations in windspeed.

10. D. Dooner and A. Seireg, *The Kinematic Geometry of Gearing: A Concurrent Engineering Approach* (New York: Wiley, 1995).

Chapter 10. Waves

1. I discuss tsunamis and other phenomena that lead to natural disasters in *Perils of a Restless Planet: Scientific Perspectives on Natural Disasters* (New York: Cambridge University Press, 1997).

2. The fact that the plane's engines are providing thrust to propel it to the west is not a counterargument. Clearly, the atmosphere itself clings to Earth, and acts to drag the plane to the east. At an airspeed of 875 mph, the plane expends its thrust simply to maintain its position in the 875-mph atmospheric flow.

3. For the details of this derivation, see R. L. Armstrong and J. D. King, *Mechanics, Waves, and Thermal Physics* (Englewood Cliffs, N.J.: Prentice-Hall, 1970), 44–50.

4. The muon was discovered in 1937 by Anderson and Neddermeyer at the California Institute of Technology, and was immediately confirmed by other groups at Tokyo and Harvard Universities.

5. G. Polya, *Induction and Analogy in Mathematics* (Princeton, N.J.: Princeton University Press, 1954).

6. See, for instance, H. Ohanian, *Modern Physics* (Englewood Cliffs, N.J.: Prentice-Hall, 1987), 180–197.

Chapter 11. Artificial and Natural Structures

1. O. G. Von Simpson, *The Gothic Cathedral: Origins of Gothic Architecture and the Medieval Concept of Order* (Princeton, N.J.: Princeton University Press, 1988).

2. Several such projects are described in *Scientific American*, Dec. 1997: C. Pelli, C. Thornton, and L. Joseph, "The World's Tallest Buildings," pp. 92–101; S. Kashima and M. Kitagawa, "The Longest Suspension Bridge," pp. 88–92.

3. M. Szpir, "Accustomed to Your Face," *American Scientist* 80 (6) (Nov. 1992): 537.

4. C. Swartz, "Justifying Atoms," *Physics Teacher* 35 (8) (Nov. 1997): 454–455.

5. I. Asimov et al., *Mysteries of Deep Space: Black Holes, Pulsars, and Quasars* (Milwaukee: Library of the Universe, 1994). See also M. C. Begelman and M. J. Rees, *Gravity's Fatal Attraction: Black Holes in the Universe* (New York: Scientific American Library, 1996).

Chapter 12. The Real and Conjectured Universe

1. S. Singh and K. A. Ribet, "Fermat's Last Stand," *Scientific American*, Nov. 1997, pp. 68–73.

2. J. Horgan, "The Intellectural Warrior," *Scientific American*, Nov. 1992, pp. 38–39.

3. L. W. Alvarez, W. Alvarez, F. Asaro et al., "Extraterrestrial Cause for the Cretaceous-Tertiary Extinction," *Science* 208 (1984): 1095–1108.

4. The Michelson-Morley experiment of the 1880s, which disproved the existence of the "luminiferous aether" is an example. See D. J. Kevles, *The Physicists* (New York: Knopf, 1978), 28–29.

5. K. Goedel and B. Meltzer (trans.), *On Formally Undecidable Propositions of Principia Mathematica and Related Systems* (New York: Dover, 1992). See also E. Nagel and J. R. Newman, *Goedel's Proof* (New York: New York University Press, 1983).

6. Bohr's statement as given by Sheldon L. Glashow, *From Alchemy to Quarks* (Pacific Grove, Calif.: Brooks/Cole, 1994), 561.

7. Also given by Glashow. The particle described here as a neutrino is today called an *antineutrino*.

8. C. G. Adler, "Neutrinos and Reality," *American Journal of Physics* 57 (1989): 878.

9. A. K. Mann, *Shadow of a Star: The Neutrino Story of Supernova 1987A* (New York: W. H. Freeman, 1997).

10. Lorenz de la Torre, "Math Is Key to Identifying Source of Strange Footprint," *Physics Today* 15 (Sept. 1997): 102.

11. Views of a number of practicing scientists on this issue can be found in J. Horgan, *The End of Science* (New York: Addison-Wesley, 1996).

INDEX

ABOUT THE AUTHOR

Dr. Ernest Zebrowski holds professorships in science and mathematics education at Southern University in Baton Rouge, Louisiana, and in physics at Pennsylvania College of Technology of the Pennsylvania State University. Prior to his academic career, he served in the Peace Corps in Liberia and held several research positions in industry. He has been recognized as an expert witness in courtroom cases involving industrial accidents, and has served on the Pennsylvania Department of Education's accrediting teams for science and mathematics teacher education programs. Three of his five books deal with applied physics. *A History of the Circle* and his recent book *Perils of a Restless Planet* (Cambridge, 1997) reflect Dr. Zebrowski's interest in rendering the processes of science and their philosophical underpinnings understandable to both nonscientists and future scientists.